RELIABILITY THEORY
AND PRACTICE

RELIABILITY THEORY AND PRACTICE

Igor Bazovsky

DOVER PUBLICATIONS, INC.
Mineola, New York

TO THE MEMORY OF MY PARENTS

Bibliographical Note

This Dover edition, first published in 2004, is an unabridged republication of the edition published by Prentice-Hall, Inc., Englewood Cliffs, New Jersey, in 1961.

Library of Congress Cataloging-in-Publication Data

Bazovsky, Igor.
    Reliability theory and practice / Igor Bazovsky.
       p. cm.
    Reprint. Originally published: Englewood Cliffs, NJ : Prentice-Hall, 1961.
    ISBN 0-486-43867-8 (pbk.)
    1. Reliability (Engineering) I. Title.

TA169.B42 2004
620'.00452—dc22

2004055125

Manufactured in the United States by Courier Corporation
43867802
www.doverpublications.com

PREFACE

I HAVE WRITTEN THIS BOOK for engineers who are confronted with the urgent problem of quantitative treatment of reliability and for students who, to become good engineers, must apply the principles of reliability in their work.

My aim was to develop reliability concepts and methods in a logical way, from simple components to complex systems, to give the reader a thorough understanding of the subject, and to show him how to solve reliability problems by analysis, design, and testing. He will find an abundance of useful reliability formulas in the book which will help him predict system reliability, establish reliability goals, and determine the procedures necessary to achieve them.

The lack of reliability wastes billions of dollars and has slowed technological progress in many vital areas. There is perhaps no engineering field today where the need for improvement is greater than in the field of reliability. I sincerely hope that this book will give many engineers a new impetus to solving reliability problems wherever they occur. Reliability is an interesting, challenging, and rewarding field once it is well understood—and when understood, it is also appreciated, because it can eliminate waste of material, time, and money.

Reliability is closely connected with the concepts of system maintainability, availability, and safety. I have therefore included a quantitative treatment of these concepts in the book and outlined the methods which have to be followed.

I hope that this work will contribute, if only in a small way, to a less wasteful, safer, and more reliable world in which the lot of humanity will improve through a concerted effort by the engineering profession.

I am indebted to Mr. Elidio J. Nucci for the encouragement he gave me to write this book. Further, I am indebted to Mrs. Julie Foster for reading the manuscript and for her useful contributions referenced in the book, and to Mr. Klaus Höhndorf for his search for applicable scientific material.

Igor Bazovsky
Chief Reliability Analyst
Raytheon Company
Missile and Space Division
Bedford, Massachusetts

TABLE OF CONTENTS

Chapter 1

THE CONCEPT OF RELIABILITY

WHAT IS RELIABILITY? Dictionaries and encyclopedias have various interpretations of this word, ranking it in the category of abstract concepts, such as goodness, beauty, honesty. Abstract concepts may mean different things to different people and are most difficult to define. What appears beautiful to one man may appear awkward to another. Concepts which are difficult to define become even more difficult to measure.

However, in engineering and in mathematical statistics, reliability has an exact meaning. Not only can it be exactly defined, but it can also be calculated, objectively evaluated, measured, tested, and even designed into a piece of engineering equipment. Thus, for us engineers, reliability is far from an abstract concept. On the contrary, reliability is a very harsh reality. It ranks on the same level with the performance of an equipment and, quite often, it is even more important than performance.

When a jet engine is designed for a certain thrust or an electronic equipment for a certain gain, and it happens in operation that the thrust or the gain is less than that called for by the specification, such performance although not the best, may under certain circumstances still be satisfactory, and the engine or the electronic equipment may turn out to be very reliable. On the other hand, an engine of a different design may supply with ease the full thrust so that it complies with all the performance specifications—but it may suddenly break down in operation. Here is where reliability enters.

Equipment breakdown can become a nightmare for the engineer—whether he is engaged in design, manufacture, maintenance, or operation. It affects not only the engineer and the manufacturer; quite often the user of the equipment bears the heaviest consequences. The price of unreliability is very high. The cure is reliability engineering.

1

From the beginning of the industrial age reliability problems have had to be considered. A classical example is ball and roller bearings; extensive studies of their life characteristics have been made since the early days of railroad transportation. Another example of the reliability approach is the design of equipment for a certain life, which dates back fifty or one hundred years. At first, reliability was confined to mechanical equipment. However, with the advent of electrification considerable effort went into making the supply of electric power reliable. Parallel operation of generators, transformers, and transmission lines and the interlinking of high-voltage supply lines into nationwide and continental grid systems have served the main purpose of keeping the supply of electric power as reliable as possible. It is not an exaggeration to say that the supply of utility electric power is nowadays almost one hundred per cent reliable. But it was far from that in the first decades of this century, as many people will remember. Parallel operation, redundancy, and better equipment have solved what was once a real problem. With the advent of aircraft came the reliability problems connected with airborne equipment, which were more difficult to solve than reliability problems of stationary or land-transportation equipment. But remarkable progress was made even in this field, mainly because of the great aircraft designers of the last few decades and their ingenious, intuitive approach.

Reliability entered a new era with the advent of the electronic age, the age of jet aircraft flying at sonic and supersonic speeds, and the age of missiles and space vehicles. Whereas originally the reliability problem had been approached by using very high safety factors which tremendously added to the weight of the equipment, or by extensive use of redundancy which again added to the over-all weight, or by learning from the failures and breakdowns of previous designs when designing new equipment and systems of a similar configuration, these approaches suddenly became impractical for the new types of airborne and electronic equipment. The terrific pace of aircraft and missile development and the miracles of modern electronics were combined with an urgent call for drastic reduction of equipment weight and size to allow the thousands and tens of thousands of necessary components to be squeezed together in small volumes. Time was running out on those who had hoped to wait to learn from mistakes made on previous designs. The next design had to be radically different from the previous one because within a few very short years the technical sciences would have again made big strides forward. Very little use could be made of the experience gained from previous mistakes; there was neither time nor money left for redesigning—both had to be made available for new projects. This rapid progress has not yet come to an end. On the contrary, the pace is increasing and will continue to increase. Therefore, the reliability problem has become more and more severe from year to year. The intuitive approach and the redesign approach have had to make way for an entirely new approach to reliability—statistically defined, calculated, and designed.

Thus, the engineer who wants to keep pace with technical developments and the manufacturer who wants to remain in business must become familiar with the new concept of reliability, and they must apply the new reliability methods in their everyday work.

Stated simply, reliability is the capability of an equipment not to break down in operation. When an equipment works well, and works whenever called upon to do the job for which it was designed, such equipment is said to be *reliable*. Satisfactory performance without breakdowns while in use and readiness to perform at the desired time are the criteria of an equipment's reliability. The equipment may be a simple device, such as a switch, a diode, or a connection, or it may be a very complex machine, such as a computer, a radar, an aircraft, a missile, or any of their subsystems. The reliability of complex equipment depends on the reliability of its components. There exists a very exact mathematical relation between the parts' reliabilities and the complex-system reliability, as we shall soon learn.

The measure of an equipment's reliability is the frequency at which failures occur in time. If there are no failures, the equipment is one hundred per cent reliable; if the failure frequency is very low, the equipment's reliability is usually still acceptable; if the failure frequency is high, the equipment is unreliable.

A well-designed, well-engineered, thoroughly tested, and properly maintained equipment should never fail in operation. However, experience shows that even the best design, manufacturing, and maintenance efforts do not completely eliminate the occurrence of failures. Reliability distinguishes three characteristic types of failures (excluding damage caused by careless handling, storing, or improper operation by the users) which may be inherent in the equipment and occur without any fault on the part of the operator.

First, there are the failures which occur early in the life of a component. They are called *early failures* and in most cases result from poor manufacturing and quality-control techniques during the production process. A few substandard specimens in a lot of otherwise fine components can easily sneak through the manufacturing process. Or, during the assembly of an equipment a poor connection may go through unnoticed. Such errors are bound to cause trouble, and the failures which then inevitably occur take place usually during the first minutes or hours of operation. Early failures can be eliminated by the so-called "debugging" or "burn-in" process. The debugging process consists of operating an equipment for a number of hours under conditions simulating actual use; when weak, substandard components fail in these early hours of the equipment's operation, they are replaced by good components; when poor solder connections or other assembly faults show up, they are corrected. Only then is the equipment released for service. The burn-in process consists of operating a lot of components under simulated conditions for a number of hours, and then using the components which survive for the assembly of the equipment.

Secondly, there are failures which are caused by wearout of parts. These occur in an equipment only if it is not properly maintained—or not maintained at all. Wearout failures are a symptom of component aging. The age at which wearout occurs differs widely with components, ranging from minutes to years. In most cases wearout failures can be prevented. For instance, in repeatedly operated equipment one method is to replace at regular intervals the accessible parts which are known to be subject to wearout, and to make the replacement intervals shorter than the mean wearout life of the parts. Or, when the parts are inaccessible, they are designed for a longer life than the intended life of the equipment. This second method is also applied to so-called "one-shot" equipment, such as missiles, which are used only once during their lifetime.

Thirdly, there are so-called "chance" failures which neither good debugging techniques nor the best maintenance practices can eliminate. These failures are caused by sudden stress accumulations beyond the design strength of the component. Chance failures occur at random intervals, irregularly and unexpectedly. No one can predict when chance failures will occur; however, they obey certain rules of collective behavior so that the frequency of their occurrence during sufficiently long periods is approximately constant. Chance failures are sometimes called "catastrophic" failures, which is inaccurate because early failures and wearout failures can be as catastrophic as chance failures, and chance failures are not necessarily "catastrophic" for the equipment in which they occur.

It is not normally easy to eliminate chance failures. However, reliability techniques have been developed which can reduce the chance of their occurrence and therefore reduce their number to a minimum within a given time interval, or even completely eliminate equipment breakdowns resulting from component chance failures.

Reliability theory and practice differentiates between early, wearout, and chance failures for two main reasons. First, each of these types of failures follows a specific statistical distribution and therefore requires a different mathematical treatment. Secondly, different methods must be used for their elimination.

Because the failure-free operation of certain equipment is vital to the preservation of human lives, to defense, and to industry, it must be highly reliable. In such equipment, early failures should be eliminated by thorough prolonged testing and check-out before it is put into service. Wearout failures should be excluded by correctly scheduled, good preventive practices. Then, if failures still occur during the operational life of the equipment, they will almost certainly be chance failures. Therefore, when such equipment is in operational use, its performance reliability is determined by the frequency of the chance failure occurrence.

Reliability engineering is concerned with eliminating early failures by observing their distribution and determining accordingly the length of the necessary debugging period and the debugging methods to be followed. Further, it is concerned with preventing wearout failures by observing the statistical distribution of wearout and determining the overhaul or preventive replacement periods for the various parts or their design life. Finally, its main attention is focused on chance failures and their prevention, reduction, or complete elimination because it is the chance failure phenomenon which most undesirably affects equipment reliability in actual operational use—in the period after the equipment has been debugged and before parts begin to wear out. For long-life equipment this amounts to the period between overhauls.

A word of caution is important here. Unfortunately, all too often not enough pains are taken to eliminate early failures completely and to prevent wearout failures. Early failures can creep into an equipment every time it is overhauled or repaired, either by an improper selection of replacement components for those that have failed and those approaching a wearout condition, or by some faulty connection (such as a solder joint), or by some other adjustments in the system which are not made properly when repair action is taken. Such poor repair practices may introduce early failures into the equipment time and again throughout its operational life; the system or equipment can never become reliable even though, with good repair practices and considering only chance failures, it might be a very reliable piece of engineering work. In a similar way wearout failures can make an inherently very reliable equipment extremely unreliable. However, such unreliability is mostly caused by negligence (for example, by not following the maintenance rules). The equipment is then not at fault.

Reliability is a yardstick of the capability of an equipment to operate without failures when put into service. Reliability predicts mathematically the equipment's behavior under expected operating conditions. More specifically, reliability expresses in numbers the chance of an equipment to operate without failure for a given length of time in an environment for which it was designed.

It is known from mathematical statistics that exact formulas exist for the frequency of occurrence of events following various kinds of statistical distributions, and from these the chance or probability of the occurrence of these events can be derived. In reliability we are concerned with events which occur in the time domain. For instance, wearout failures usually cluster around the mean wearout life of components. Once their distribution is known, the probability of wearout failure occurrence at any operating age of the component can be mathematically calculated. Similar considerations apply to early failures and to chance failures. However, early and chance failures do not cluster around any mean life but occur at random intervals. They therefore belong in the category of random events or stochastic processes and have their own characteristic distribution which is different from wearout failures. Although the time of occurrence of failures which occur at random time intervals cannot be predicted, the probability of the occurrence or nonoccurrence of such failures in an operating interval of a given length can be calculated by means of the theory of probability.

By its most primitive definition, *reliability* is the probability that no failure will occur in a given time interval of operation. This time interval may be a single operation, such as a mission, or a number of consecutive operations or missions. The opposite of reliability is *unreliability*, which is defined as the probability of failure in the same time interval.

The word *probability* is sometimes accepted with an amount of skepticism, especially in an exact science like engineering. In fact, it is not quite self-evident what the mathematical concept of probability has to do with the design or operation of an engineered equipment. Even if it could be applied by some tricky calculations, how could any confidence result from such an unorthodox approach when engineering and physics are already such exact sciences? Such reasoning would be very self-deceptive. A closer look into the science of physics reveals immediately that numerous physical phenomena can be exactly described only with the help of the theory of probability and that probability calculations have a very realistic meaning. There are very important areas in physics, especially in molecular, atomic, and nuclear physics, which can mathematically be formulated only with the help of the theory of statistics and probability. It is the variability of the natural world which compels us to use the probability approach to these problems. Two widely known examples from physics which are of fundamental importance for the behavior of all matter are the decay law of radioactive substances, and statistical mechanics, which deals with the change of state of atoms, electrons, and other basic particles. The engineer is aware that the equipment he designs and builds consists of molecules and atoms, which in turn are built from neutrons, protons, and electrons. He should not be surprised to find that equipment failures which result from interactions of heat, electric and magnetic fields, and mechanical vibrations can best be mathematically described by the same probability laws which so exactly describe things such as the atomic disintegration of radioactive substances. Further, the engineer knows from his own experience that once a problem can be expressed closely enough in the language of mathematics, formulas will result which can lead to the solution of an engineering problem.

Of course, one may wonder how exact probability calculations are, or how exact can they be. Can they give answers which are close enough to reality? A simple example, that of coin tossing, is very revealing in this respect. The probability of obtaining a head when tossing a well-balanced coin once is $\frac{1}{2}$ or 50 per cent, as everybody knows. The probability of obtaining a tail is also $\frac{1}{2}$. Therefore the probability of obtaining a head or a tail in a single throw is $\frac{1}{2} + \frac{1}{2} = 1$, or 100 per cent. A probability of 100 per cent is certainty. Similarly a probability of 0 per cent is impossibility in the sense that the event cannot occur. When the probability of obtaining a head or a tail in a single throw equals unity, then the probability that we obtain neither a head nor a tail in a single throw must be $\frac{1}{2} - \frac{1}{2} = 0$. Thus the event of obtaining neither a head nor a tail in a single throw is not possible, and therefore its probability is 0 per cent. We have discounted here the possibility that the toss would end with the coin standing on its edge. We have simplified the calculation by admitting only two outcomes—head or tail. This is, however, a very good assumption and is therefore permissible. We use this approach of two mutually exclusive events—head or tail—very often in reliability calculations where we deal with the concepts of "success or failure," and discount the possibility of a third outcome.

Working the coin example further, it is correct to say that when the coin is tossed several times the probability of obtaining a head in each throw is $\frac{1}{2}$ every time, regardless of how many times the coin has already been tossed. So we would most likely expect to obtain 5 heads and 5 tails if we tossed the coin 10 times. However, here again the variability of nature comes into play. It is quite unlikely that we would obtain such a result in a single series of 10 tosses. The experiment may easily end with 3 heads and 7 tails, or any other result. Still, it would be incorrect to conclude from such a result that the probability of the coin yielding a head in a single throw is not $\frac{1}{2}$. It would be wrong to conclude from the first 10 tosses that the probability of obtaining a head is $\frac{3}{10} = 0.3$ merely because we happened to obtain 3 heads in 10 tosses.

We now have arrived at a very important point which must be carefully considered in reliability engineering, especially in reliability testing. When the coin experiment is repeated again and again, and the outcomes of each of the successive series of 10 tosses are added up, we shall find that the larger the number of tosses, the closer we come to the result predicted by the coin's probability of $\frac{1}{2}$. For instance, out of 100 tosses we may find that there were 42 heads and 58 tails. So, we would now calculate the probability of the coin yielding heads to be $^{42}/_{100} = 0.42$. This brings us closer to $\frac{1}{2}$ than the first result of 0.3. And if we were to continue the experiment several hundred times, we would come very close to $\frac{1}{2}$. It is customary to call $\frac{1}{2}$ the *true probability*, or simply the *probability* of the coin, and the result computed from a number of trials, for instance 0.3 and 0.42 in our coin example, the *estimate of probability*.

A few important conclusions, which also apply to reliability, can be drawn from the coin experiment:

1. The coin has a true probability for yielding heads when tossed. When the coin is unbiased, this true probability is 0.5.
2. An estimate of the true probability can be obtained from a number of trials. In the case of the coin we had 0.3 from 10 tosses and 0.42 from 100 tosses.
3. The larger the number of trials, the more likely it is that we are getting very close to the true probability.
4. When the true probability of an event is not known, there exists an experimental way to estimate it—by performing a number of trials.
5. When the number of trials is small, the estimate may be quite off from the true value. It may be overpessimistic or overoptimistic, but it also may be very close to the true value. Estimates obtained from a small number of trials have to be taken very cautiously.

The number of trials from which an estimate of probability is obtained is an exact measure of the likelihood of how close to the true value the estimate is. Statistical methods have been developed which enable us to calculate, from the obtained estimate and from the known number of trials, confidence limits for the true probability and the confidence in per cent that the true probability lies within these limits.

The mathematical relation between the true probability of an event and an estimate of that probability obtained from N trials is a limit function. If in N trials we obtain n outcomes which yield the event whose probability we wish to estimate, the probability estimate is defined as

$$P_{est} = \frac{n}{N} \tag{1.1}$$

It is customary to use the symbol $\hat{P}$ (P hat) for an estimate of probability. The true probability P is then defined as the limit:

$$P = \lim_{N \to \infty} \hat{P} = \lim_{N \to \infty} \frac{n}{N} \tag{1.2}$$

From this results the following definition: True probability is the limit of the ratio of favorable outcomes to the total number of trials as the total number of trials approaches infinity.

The consequence of this definition of true probability is that when it is not known *a priori*, we can never learn its exact value because we can never have an infinite number of trials from which to compute it. All we can obtain are better or worse estimates, depending on the number of trials. It is therefore a theoretical matter to distinguish between true probabilities and probability estimates. Thus, we shall use the symbol P without a hat for probability, and only in cases where we are particularly interested in confidence limits shall we differentiate between the true and estimated values. But we shall always bear in mind that when our probability figures are derived from small samples, they may be only poor estimates (though not necessarily).

Projected into the field of reliability, we find that reliability calculations, being probability calculations, apply to idealized models. For instance, if it is known from a large number of tests that the reliability of a system to perform an exactly defined operation is 0.9 or very close to 0.9, it does not necessarily follow that in 10 operations the system will perform satisfactorily exactly 9 times and fail once. It may fail two or three times in 10 trials, and it may not fail at all in 10 trials. But in a large number of trials it will perform well in about 90 per cent of the trials and will fail in about 10 per cent of the trials. So, we can imagine an ideal model of this system which would operate without failure exactly 9 times in 10 operations. Using this model, we would expect that out of 1000 operations, 900 would be completed successfully and 100 would end with a failure. And if our estimate of 0.9 reliability was a good estimate, we shall find that the real system in 1000 operations will yield results which come very close to the idealized model. We can then safely say that the real system has a probability of 0.9 to complete every individual operation. However, even for the idealized model it is impossible to predict in which of ten successive operations the expected one failure will occur, just as it is impossible to predict in which throw a coin will fall head up. The nature of probability is such that it allows us to predict quite closely the number of certain events in a large number of trials, but the outcome of a single trial cannot be predicted. Only the chance of the event happening in the single trial can be very closely estimated by probability calculations. Such calculations are performed by means of very exact mathematical theorems of the probability calculus. We shall make extensive use of these theorems, especially the theorem of addition of probabilities, the theorem of the multiplication of probabilities, and Bayes' theorem. We assume that the reader has a knowledge of these theorems,* as well as a knowledge of the fundamentals of mathematical

---

* William Feller, *An Introduction to Probability Theory and Its Applications*, John Wiley & Sons, Inc., New York, 1957.

statistics.* However, we shall endeavor here to make it as easy as possible for the reader to follow the development of statistics and probability calculations as they are applied to the theory of reliability.

As we said before, the reliability of an equipment is the probability that it will not fail to operate satisfactorily in a given time interval of length t. This probability can be estimated by means of very exact statistical and probability calculations. The estimate can come very close to the true reliability of the equipment. When comparing the estimates with actual experience gained from the operation of the equipment, we find that the calculated figures correlate well with experience for a large number of operations, but that the correlation is usually poor when only small numbers of trials are considered. When related to observed results, reliability figures derived by calculations apply to averages computed from a significant number of observations. To return briefly to the example of an unbiased coin with a probability of $\frac{1}{2}$, when we compute the outcome of 10 observed tosses it may appear that $\frac{1}{2}$ is not the exact probability figure. But when the results of 100 or 1000 tosses are counted, we shall soon agree that $\frac{1}{2}$ is the best figure after all and that the use of any other figure which we may have found from the first few tosses would only lead us farther from reality.

* A. Hald, *Statistical Theory with Engineering Applications*, John Wiley & Sons, Inc., New York, 1952

Chapter 2

THE DEFINITION OF RELIABILITY

To ESTIMATE RELIABILITY and to perform reliability calcultions, we must first define reliability. It is not enough to say that it is a probability. A probability of what? When we say that a coin has a probability of 0.5, this only makes sense when we add that this is the coin's probability of falling head up, or falling tail up. It is not the coin's probability of standing on edge after a toss (the probability of this is assumed to be zero), nor is it the coin's probability of turning in the air a given number of times. With a probability figure there is always an associated, exactly defined event.

In its simplest and most general form, reliability is a probability of success. A widely accepted definition reads: "Reliability is the probability of a device performing its purpose adequately for the period of time intended under the operating conditions encountered."* This definition implies that reliability is the probability that the device will not fail to perform as required for a certain length of time. Such probability is also referred to as the *probability of survival*. In most cases the probability with which a device will perform its function is not known at first. It is not identical with the case of an unbiased coin, where the probability of the coin to yield heads or tails is known and the statement that this probability is ½ can be made with 100 per cent confidence. Actually, the case of reliability can be compared with a biased coin which has a very marked tendency to fall head up rather than tail up. Such a coin will have a definite probability of yielding heads, but its actual value will not be known. We can estimate this probability by performing a

* C. R. Knight, E. R. Jervis, and G. R. Herd, *Terms of Interest in the Study of Reliability*, ARINC Monograph No. 2, Aeronautical Radio, Inc., Washington, May 25, 1955.

11

number of tosses and statistically evaluating the observed outcomes. The larger the number of tosses, the closer our estimate will come to the true probability, and the more confidence we can have in our estimate. This is the kind of problem we face in reliability studies. The true reliability is never exactly known, which means the exact numerical value of the probability of adequate performance is not known. But numerical estimates quite close to this value can be obtained by the use of statistical methods and probability calculations. How close the statistically estimated reliability comes to the true reliability depends on the amount of testing, the completeness of field service reporting all successes and failures, and other essential operational data. When reliability estimates are computed by means of the probability calculus, which can be done in the case of complex equipment, the accuracy of the estimate depends on how well we know the reliabilities of the components from which the equipment is composed for the given operating conditions. For the statistical evaluation of an equipment (i.e., of the probability that the equipment will not fail to perform adequately a specified function), the equipment has to be operated and its performance observed for a specified time under actual operating conditions in the field or under well-simulated conditions in a laboratory. The event whose probability is to be estimated is the adequate performance of a specified function for a length of time. Criteria of what is considered an "adequate performance" have to be exactly spelled out for each case, in advance. If laboratory testing is used, the operating and environmental conditions must be exactly specified also, because the reliability estimate obtained will refer only to the spelled-out performance requirements and to the operating conditions and the environment in which the equipment operates during the test. The spelling out of the function (in terms of an exact excerpt from the performance specification or in terms of the required performance) which the equipment must perform when put into operation is imperative because only this method will catch all cases of failure, and only in this case can the criteria of failure be specified.

Measurement of the adequate performance of a device requires measuring all important performance parameters. As long as these parameters remain within the specified limits, the equipment is judged as operating "satisfactorily." When the performance parameters drift out of the specified tolerance limits, the equipment is judged as having malfunctioned or failed. For instance, if a system of four 30-kva generators in an airplane is capable of supplying 120 kva at $115\frac{5}{200}$ volts and 400 cps, but the total electrical load requirements are not more than 60 kva, the generating system will not fail if one or two generators fail in flight. The satisfactory performance of the generating system is specified in this case as the system's capability to supply 60 kva at $115\frac{5}{200}$ volts and 400 cps, and system failure occurs only when the four generators are not capable of supplying the 60 kva of the required quality. Therefore, two generators may fail and the system will still operate satisfactorily. In other words, no system failure occurs in spite of the failure of two generators.

It is immediately obvious that if the total continuous load is increased to 70 kva, the system will be considerably less reliable than for a 60-kva load, although nothing has changed in the generating system itself. Now only one generator is allowed to fail, because the failure of two generators means that the system cannot supply 70 kva, but only 60 kva. Some loads would have to be disconnected in order not to overload the generators, and therefore the system would no longer operate "adequately." We are actually interested in several reliabilities of such a system. For instance, in the case of four generators and a total 60-kva load, we may want to know the system reliability to supply only these 60 kva. We may also want to know the system reliability to supply only the essential loads amounting to, say 25 kva, and finally, we may want to know the system reliability to supply an absolute minimum of electrical power in case of emergency. The system's reliability may be different in all three cases; in fact, it usually is. The precise definition of what is understood under "adequate performance" is all important in reliability studies, because we are seeking the probability with which the equipment, while in operation, will satisfactorily fulfill a required function.

In the probability context, satisfactory performance is directly connected to the concepts of failure or malfunction. The relation between the two is one of mutually exclusive events—which means the equipment, when in operation, is either operating satisfactorily or has failed or malfunctioned. Sometimes it may be simpler to specify first what is regarded as "failure" or "malfunction," and satisfactory performance is then every other operating condition which is not a failure or malfunction. If this clear-cut distinction is not made and some third condition in the equipment's operation (besides satisfactory performance and malfunction or failure) is included, it will become necessary to evaluate more than one reliability because there will be at least three probabilities involved: one probability for what is termed satisfactory performance, one for malfunction or failure, and one for the occurrence of the third condition. Such an approach can and should be avoided because it complicates the evaluation, especially in reliability tests. If several probabilities must be evaluated for one system, the simplest approach is to split the problem into several individual problems and deal with each separately by using the principle of two mutually exclusive events, that is, events which mutually exclude each other, such as *success–failure*.

Once adequate performance (i.e., *success*) has been clearly defined for a specific case, then the evaluation of reliability can begin. To judge

adequate performance involves the observation of inadequate performance in operation, therefore, the observation of malfunctions and failures
which violate the requirement for an "adequate performance." The frequency at which malfunctions and failures occur is then used as a parameter for the mathematical formulation of reliability. This parameter is
called the *failure rate*. It is usually measured in number of failures per
unit operating hour. Its reciprocal value is called the *mean time between
failures* and this is measured in hours.

We have placed great emphasis on the need for a clear-cut definition
of the function of a device and its adequate performance on the one hand,
and of failure or malfunction on the other. At first glance the need for
this emphasis may seem to be exaggerated, but in later chapters it will
become well understood.

It is true that in some simple cases, where devices of the "go—no go"
type are involved, the distinction between adequate performance and
malfunction is a very simple matter. For instance, a switch either works
or does not work—it is good or bad. In such cases it is simple to define
the function (the closing or opening of a circuit when the switch is actu-
ated) and to define what is understood under "adequate performance"
on the one side, and under "malfunction" or "failure" on the other. But
there are many more cases of a nature such that a clear-cut decision cannot
be made so easily and a number of performance parameters and their
limits must first be specified; operation within the limits is considered
adequate or satisfactory, and outside of the specified limits it is considered inadequate.

Devices requiring a clear-cut and spelled-out definition of adequate
performance in terms of parameters and limits are capacitors, tubes, and
other electronic components. Also included are all devices with specified
performance tolerances, and finally, most systems. For instance, when
an aircraft electrical generating system of several generators is required
to supply 60 kva at a given voltage, waveshape, and frequency, but suddenly is capable of supplying only 40 kva, this system condition will have
to be regarded as system failure from the viewpoint of some loads but as
adequate performance for other loads. Similar differences must be carefully considered when evaluating the reliability of systems. This is the
case when several reliabilities of a system have to be computed. We shall
return to this problem later.

When failures occur at random during the operation of a device, the
failure rate, which can be derived from the number of failures observed
in an operating time of sufficient length, allows one to calculate the reliability of the equipment by means of a simple mathematical formula with
which we shall become closely acquainted in the following chapter. Fail-

ure rates of devices are obtained experimentally either from reliability tests or from observations during in-service operation.

The term "adequate performance" in itself does not define sufficiently the purpose of calculating the probability of success of complex operations. The functions of even simple devices are frequently quite complicated processes. Usually a number of parts must operate properly to perform the function of the device. When an equipment has numerous components, the complexity of functions which the individual components have to perform makes the problem very involved. If all parts operate properly, the equipment usually will also operate properly. But even this is not always true. There are cases when all parts operate within their individually specified performance limits, but the equipment may operate unsatisfactorily if the performance of several components has drifted close to their limits in such a way that the combined effect of the drifts causes the equipment to operate outside the equipment performance limits. In such cases, no individual component can be said to have failed but the equipment has failed. These failures occur most frequently in electronic systems. The problem is one of electronic engineering more than one of reliability, for it requires proper electronic system design to allow for multiple component drifts. Conversely, there are situations when a part fails and the equipment still operates satisfactorily. This applies especially to cases of redundancy. Redundant components, networks, etc., are included in an equipment for the specific purpose of insuring satisfactory operation in case of partial failures.

Since reliability is a yardstick of capability to perform within required limits when in operation, it normally involves a parameter which measures time. This may be any time unit, which is preferable in cases where continuous operation is involved; it may be a number of cycles when the equipment operates only sporadically, in regular or irregular periods, or a combination of both. It is meaningful to speak of the operating hours of an engine, generator, aircraft, etc., and to calculate the probability that no failure will occur in a given number of hours of operation. But for a switch or relay it may be more meaningful to speak of the number of cycles or number of operations which such a device has to perform. The probability that no failure will occur in a number of operations (cycles) may in these cases tell much more than the probability of no failure in a number of hours. The lifetime of a switch is more plausibly expressed in the number of switchings than in the number of hours. Thus, a switch measures its "time" in cycles of operation rather than in hours. However, it is often possible to correlate the number of cycles with hours of operation, especially when some regularity in the cycles exists. For instance, when it is known that on an average a switch operates 5 times in 10 hours, its reliability expressed in terms of 5 switching operations or in terms of 10 operating hours may result in the same numerical probability value. But the same switch, when used in another system where it performs 100 switching operations in 10 operating hours, will have a much lower reliability for 10 hours than in the first case. The number of hours remains the same but the number of cycles has changed. Thus it is necessary, whenever a correlation between straight time and other life parameters (e.g., cycles, or number of revolutions) is sought, to find the correlating figure for each individual application. The conversion of cycles into hours may be very useful in complex probability calculations. Such conversion brings the component lives onto a common parameter and greatly simplifies the mathematical operations. The role of the parameter "time," or its equivalent, will become clear in the following chapter where the basic concept of reliability is formulated mathematically.

Chapter 3

THE EXPONENTIAL CASE OF
CHANCE FAILURES

I N THE SIMPLEST CASE, when a device is subject only to failures which occur at random intervals, and the expected number of failures is the same for equally long operating periods, its reliability is mathematically defined by the well-known exponential formula

$$R(t) = e^{-\lambda t} \tag{3.1}$$

In this formula e is the base of the natural logarithm $(2.71828\ldots)$, λ is a constant called the *chance failure rate*, and t is an arbitrary operating time for which we want to know the reliability R of the device. The failure rate must be expressed in the same time units as time, t—usually in hours. The reliability R is then the probability that the device, which has a constant failure rate λ, will not fail in the given operating time t.

This reliability formula is correct for all properly debugged devices which are not subject to early failures, and which have not yet suffered any degree of wearout damage or performance degradation because of their age. The life period of the device in which the above formula is valid is conventionally referred to as the "useful life" of the device. Its length varies widely for different devices. It is important that the time t in the formula $R(t) = e^{-\lambda t}$ never exceed the useful life of the device.

For instance, if a component has a useful life of only 1000 hours, by means of Equation (3.1) we can predict its reliability for any operating time interval within these first 1000 hours of the component's life. However, the formula will give a wrong answer if any operating period beyond the useful life is considered, because after the component has exceeded its useful life period, the failure rate will begin to increase steadily. In the useful life period, the reliability of a device is approximately the same for operating times of equal length. Thus, for the first 10 hours of useful life the reliability is the same as for the last 10 hours, i.e., the reliability is the same for a 10-hour operation from 0 to 10 hours and from 990 to 1000 hours, if the device has survived up to 990 hours. We shall qualify this statement later.

To illustrate the important fact of an equal chance of survival for periods of equal length throughout the useful life, let us assume that a debugged device with a 1000-hour useful life has a constant failure rate of $\lambda = 0.0001$ per hour in a given environment. Its reliability for any 10 hours' operation within these 1000 hours is

$$R = e^{-0.0001 \times 10} = e^{-0.001} = 0.9990$$

or 99.9 per cent.*

The probability that the device will not fail in its entire useful life period of 1000 hours is

$$R = e^{-0.0001 \times 1000} = e^{-0.1} = 0.9048$$

Thus, it has a chance of 90 per cent to survive up to 1000 hours, counted from the moment when first put into operation. But if it survives up to 990, then its chance to survive the last 10 hours (from 990 to 1000 hours) of its useful life is again 99.9 per cent, or $R = 0.999$.

However, if this device should continue operation beyond a total of 1000 hours, wearout will begin to play a role and the reliability of the device will decrease for each subsequent 10-hour operation, sometimes quite rapidly, because the failure rate will be increasing. Thus, we see that the reliability of a debugged device is constant for operating time intervals of equal length within its useful life period, given that the device has survived the preceding operations; but the reliability of the device will decrease more or less rapidly if it has reached the end of its useful life period.

The first part of this statement can also be expressed as follows: The reliability of a device, or the probability that the device will not fail within a given time interval, is a simple exponential function of that time interval, given that the device has survived to the beginning of the time interval and provided that the age of the device is such that it does not reach the end of its useful life within the time interval under consideration. To indicate that reliability is a function of the time interval under consideration, we write $R(t) = e^{-\lambda t}$ and we can plot, for a given environ-

---

* The value of R for a given exponent $-\lambda t = -x$ is obtained from tables such as *Tables of the Exponential Function e^x*, National Bureau of Standards, Applied Mathematics Series No. 14, U.S. Government Printing Office, Washington, D.C., June 28, 1951.

ment, the reliability function $R(t)$ in the form of a graph for any device which has a constant and known failure rate in that environment. The parameter λ completely determines the reliability of the device.

For reasons which will become obvious later, we often use the reciprocal value of the failure rate, which is called the *mean time between failures, m*. The mean time between failures, abbreviated MTBF, can be measured directly in hours. By definition, in the exponential case, the mean time between failures, or MTBF, is

$$m = \frac{1}{\lambda} \tag{3.2}$$

It is a time parameter and, in the same way as λ, it completely describes the reliability of an exponentially behaving device or system. The reliability function $R(t)$, also called *probability of survival function*, can therefore also be written in the form

$$R(t) = e^{-t/m} \tag{3.3}$$

When plotting this function, with R values on the ordinate and the corresponding t values on the abscissa, we obtain a curve which is often referred to as the *survival characteristic* and is shown in Figure 3.1. Let us discuss this graph.

First, it is important to understand that the time t on the abscissa is not a measure of the calendar life or of the total accumulated operating life of the device since it was new. It counts only the hours of any arbitrarily chosen operating period with $t = 0$ designating the beginning of the considered operating period, not calendar zero time when the device enters service for the first time. To distinguish between the total accumulated operating life of a device and its operating time in a given interval, it would actually be better to use T for the former and t for the latter. However, we can use t for both as long as we clearly differentiate between them.

Thus, the time t in the formula $R(t) = e^{-\lambda t} = e^{-t/m}$ measures the operating hours (or other time units) in an arbitrarily chosen operating period of a device, regardless of how many hours the device has already been in operation before this specific operating period. It serves to predict the chance of survival (R) of the device for any period within the useful life of the device, i.e., for a "mission" of a certain length t, where t is counted from the time $t = 0$ when the mission begins. Therefore, t is often called *mission time*. It is assumed that the device has survived previous missions, and it will not reach the end of its useful life in the mission now under consideration. The first assumption is written as $R = 1$ at $t = 0$, which means that the device has survived to the beginning of the mission. The second assumption is contained in the original assumption of $\lambda =$

constant, on which the formula

$$R(t) = e^{-\lambda t}$$

is based.

Second, it is seen that the time t in the graph extends to infinity, which seems to make no sense. However, when only chance failures are

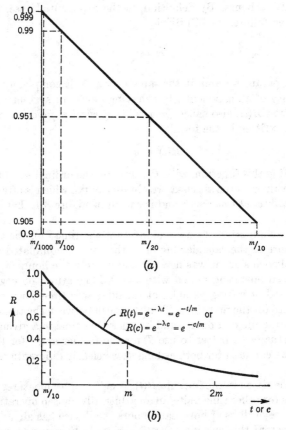

Fig. 3.1. The standardized reliability curve. (*a*) Upper portion of the reliability curve. (*b*) The curve.

considered, the certainty that a device will fail because of a chance failure exists only for an infinitely long operating period. Conversely, the probability that the device will not fail because of chance becomes zero only as t approaches infinity. Clearly, these statements do not take wearout into account. Naturally, if a device does not fail of chance, sooner or later it will fail because of wearout. But we already have specified that we consider here only chance failures, which means we determine not to operate the device beyond its useful life. Within its useful life the device is always as good as new because its failure rate remains the same; wearout has had no time to cause such damage to the device that it would be more susceptible to failure. Consequently, the probability that the device will fail by chance remains the same for periods of equal length throughout the entire useful life. If we had an ideal device which would never wear out, such device could fail only by chance, and its probability of failure would become 100 per cent only as t approached infinity.

In practical reliability studies we are usually concerned with predicting the probability of survival for a given mission length. These mission times are almost always much shorter than the useful life of a device and also much shorter than its mean time between failures. Therefore, reliability predictions are usually made only for time intervals which correspond to the extreme upper portion of the reliability curve.

There are a few points on this curve which are easy to remember and which help greatly in rough predicting work. For an operating time $t = m$, the device has a probability of only 36.8 per cent (or approximately 37 per cent) to survive. Its reliability for a mission time of $t = m$ hours is only 0.368. In practice, this means that if we were to operate 100 components of the same kind, after $t = m$ hours only about 37 would be still operating, whereas about 63 would have failed before that time.

For $t = m/10$, the curve shows a reliability of $R = 0.9$, or approximately 90 per cent. For $t = m/100$, the reliability is $R = 0.99$; for $t = m/1000$, it is 0.999. A reliability of 0.999 means that out of 1000 equal components operated for $t = m/1000$ hours, we would expect 999 to survive and one to fail. Going further on the reliability curve, we find that for an operating time of $t = m/10,000$ the reliability becomes approximately 0.9999; for $t = m/100,000$, it becomes 0.99999; for $t = m/1,000,000$, it becomes 0.999999.

These points on the reliability curve apply to any component and to any system which behaves exponentially. Therefore, a reliability curve which uses m as a time unit is a standardized curve. In the case of a component, m and R in the curve are the mean time between failures and the reliability of that component. In the case of a system, m and R are the mean time between failures and the reliability of that system.

Thus, if a component is required to have a reliability of 0.999999 for 1 hour of operation, it must have a mean time between failures of 1,000,000 hours. If a system is required to have a reliability of 0.9999 for 1 hour of operation, it must have a mean time between failures of 10,000 hours. For 10 hours of operation, the same system would then have a reliability of approximately 0.999, for 100 hours of operation, approximately 0.99.

There exists a definite relationship among component reliabilities, the number of components in a system, and system reliability. If a component has a reliability of 0.999 for $t = m/1000$ hours, this means that out of 1000 of these components operated for $t = m/1000$ hours, we would expect one to fail and 999 to survive. The chance that all 1000 components would survive an operating time of $m/1000$ hours is only 0.368. Therefore, a system composed of 1000 of these components would have a reliability of only 0.368 to survive an operation of $t = m/1000$ hours, if the failure of any component causes the system to fail. Thus the system's mean time between failures is only $m_{system} = m/1000$ hours. But if the system were to be built of only 100 of these same components, the chance that one or more of these components would fail in $t = m/1000$ hours is only about $100/1000 = 0.1$; therefore the system would have a reliability of approximately 0.9 to survive for $t = m/1000$ hours. And if we had only 10 components in the system, the chance that one or more of them would fail in $m/1000$ hours would be only about $10/1000 = 0.01$; system reliability would therefore be about 0.99 for $t = m/1000$ operating hours. This corresponds to a system mean time between failures of $m/10$. It follows that a system composed of n equal components, each of which has a mean time between failures of m hours, has a system mean time between failures of m/n hours, if the failure of any component causes system failure. This applies also to unequal components when m is the average mean time between failures of the components. We shall return later to a detailed treatment of system reliability calculations from component reliabilities.

Finally, it should be mentioned that m does not need to be given in time units. In certain cases it is more meaningful to change the time units to other parameters which characterize an equipment's operational life better than time units. For instance, in the case of switching devices (switches, relays, etc.), the number of operating cycles is a more appropriate measure of life than the number of operating hours. In such cases, the time scale on the abscissa of the reliability curve is changed to a cycle scale, and m becomes the mean number of cycles between failures. Its reciprocal is then the failure rate per one operating cycle.

Thus a switch which has a mean number of cycles between failures m has a reliability of 0.99 for $m/100$ cycles and a reliability of 0.999 for $m/1000$ cycles, etc. The reliability equations (3.1) and (3.3) can then be rewritten substituting the number of cycles c for the time t:

$$R(c) = e^{-\lambda c} = e^{-c/m} \qquad (3.4)$$

In this equation c is the number of operating cycles for which we want to calculate the reliability of the switching device, m is the mean number of cycles between failures of the device, and λ is the failure rate of the device

expressed in failures per one cycle. The reliability of the switching device
for a single one-cycle operation is then*

$$R(1) = (0.368)^\lambda \tag{3.5}$$

which can be approximated to $R(1) = 1 - \lambda$ for small failure rates λ,
as we shall see later. This reliability for a single switching operation is
often used in system reliability calculations where a switch or a relay has
to switch only once during an entire mission of the system—for instance,
to close a circuit at the beginning of the system's operation.

When a switch performs a given number of operations during a system mission time t, it is possible to relate the number of operating cycles c
to the number of operating hours t, and to express the reliability of the
switch in terms of the system operating time t. In complex reliability
calculations this approach can be used with advantage to simplify these
calculations.

In this chapter we have shown that reliability can be formulated
mathematically, and that in the simplest case the formula yields a characteristic exponential survival curve. Further, we have related the failure
rate to the mean time between failures in the exponential case, and have
derived some easy-to-remember reliability values for mission times equal
to simple decimal fractions of the mean time between failures. We have
also related the component mean time between failures to the system
mean time between failures for the exponential case, and have shown
that in some instances it is more meaningful to use the mean number of
cycles between failures to describe the reliability of certain devices.

* Note that $e^{-\lambda} = (1/e)^\lambda = (0.368)^\lambda$.

Chapter 4

THE DERIVATION OF THE
RELIABILITY FUNCTION

IN CHAPTER THREE the exponential formula $R(t) = e^{-\lambda t}$ was used to calculate the reliability of a device, equipment, or system which has a constant failure rate. In this chapter it will be shown that the exponential formula can be derived mathematically from the basic definition of probability. The mathematical derivation which now follows will result in a general reliability formula which is always valid, regardless of whether the failure rate is constant or variable, and the formula $R(t) = e^{-\lambda t}$ is just a special case of the general formula.

To make this derivation we start from the very fundamental definition of probability. The probability of an event A is defined as the fraction of the favorable outcomes of the event A to the total number of trials in a test, if each trial has an equal chance to result in the event A. Thus, when each trial ends either with a favorable outcome of the event A or with an unfavorable outcome, which we denote as event B, and there are X outcomes with the attribute A, and Y outcomes with the attribute B, then the total number of trials is $X + Y$, and the probability of A is

$$P(A) = \frac{X}{X + Y} \tag{4.1}$$

Equally, the probability of B will be defined as

$$P(B) = \frac{Y}{X + Y} \tag{4.2}$$

Strictly speaking, these probabilities are only approximations of the true probabilities $P(A)$ and $P(B)$. Their exact values could be obtained only from an infinite number of trials. But when the number of trials $X + Y$ is reasonably large, the estimate will be very close to the true probability. When event A is the survival of a component and event B is its failure, then using the above definition of probability, the reliability of a component can be defined as the fraction of components surviving a test to the total number of components present at the beginning of the test.

When a fixed number N_0 of components are repeatedly tested, there will be, after a time t, N_s components which survive the test and N_f components which fail. Therefore $N_0 = N_s + N_f$ is a constant throughout the test because as the test proceeds, the number of failed components N_f increases exactly as the number of surviving components N_s decreases. The reliability or probability of survival, expressed as a fraction by the probability definition, is at any time t during the test

$$R(t) = \frac{N_s}{N_0} = \frac{N_s}{N_s + N_f} \qquad (4.3)$$

where N_s or N_f are counted at that specific time t. Thus, the fraction tells us the probability of nonfailure of any single component if operated for the time t. Obviously, as the test proceeds and N_s decreases because more and more components fail, the probability of survival, or reliability, also decreases proportionally. Thus, reliability measured in such a test is a function of the operating time t. In the same way as we defined reliability, we can also define the probability of failure Q (called unreliability) as

$$Q(t) = \frac{N_f}{N_0} = \frac{N_f}{N_s + N_f} \qquad (4.4)$$

It is at once evident that at any time t,

$$R + Q = 1$$

because $(N_s + N_f)/(N_s + N_f) = 1$. The events of component survival and component failure are called *complementary* events because each component will either survive or fail. They are also called *mutually exclusive* events because if a component has failed, it has not survived, and vice versa.

The number of surviving components in a test is $N_s = N_0 - N_f$; therefore reliability can also be written as

$$R(t) = \frac{N_s}{N_s + N_f} = \frac{N_0 - N_f}{N_0} = 1 - \frac{N_f}{N_0} \qquad (4.5)$$

By differention of this equation we obtain

$$\frac{dR}{dt} = \frac{d(1 - N_f/N_0)}{dt} = -\frac{1}{N_0}\frac{dN_f}{dt} \qquad (4.6)$$

because N_0 is constant. Rearranging (4.6) we get

$$\frac{dN_f}{dt} = -N_0 \frac{dR}{dt} \tag{4.7}$$

which is the rate at which components fail. But because $N_f = N_0 - N_s$ and

$$\frac{dN_f}{dt} = \frac{d(N_0 - N_s)}{dt} = -\frac{dN_s}{dt}$$

Eq. (4.7) is also the negative rate at which components survive.

The term dN_f/dt can also be interpreted as the number of components failing in the time interval dt between the times t and $t + dt$, which is equivalent to the rate at which the component population still in test at time t is failing.

At the time t we still have N_s components in test; therefore dN_f/dt components will fail out of these N_s components. When we now divide both sides of Equation (4.7) by N_s, we obtain on the left the rate of failure or the instantaneous probability of failure per one component, which we call the failure rate λ:

$$\lambda = \frac{1}{N_s} \frac{dN_f}{dt} = -\frac{N_0}{N_s} \frac{dR}{dt} \tag{4.8}$$

But from $R = N_s/N_0$ we know that $N_0/N_s = 1/R$ and substituting this on the right side of (4.8) we get

$$\lambda = -\frac{1}{R} \frac{dR}{dt} \tag{4.9}$$

which is the most general expression for the failure rate because it applies to exponential as well as nonexponential reliabilities. In the general case, λ is a function of the operating time t, for both R and dR/dt are functions of t. Only in one case will the equation yield a constant, and that is when failures occur exponentially at random intervals in time. By rearrangement and integration of (4.9), we obtain the general formula for reliability,

$$\lambda \, dt = -\frac{dR}{R}$$

$$\int_0^t \lambda \, dt = -\int_1^R \frac{dR}{R} = -\ln R$$

$$\ln R = -\int_0^t \lambda \, dt$$

Solving for R and knowing that at $t = 0$, $R = 1$, we obtain

$$R(t) = e^{-\int_0^t \lambda \, dt} = \exp\left[-\int_0^t \lambda \, dt \right] \tag{4.10}$$

So far in this derivation we have made no assumption regarding the failure rate λ, and therefore λ can be any variable and integrable function of the time t. Consequently, in the equation $R(t)$ mathematically describes reliability in a most general way and applies to all possible kinds of failure distributions.

When we specify that λ is constant in Equation (4.10), the exponent becomes

$$- \int_0^t \lambda \, dt = -\lambda t$$

and the known reliability formula for constant failure rate results,

$$R(t) = e^{-\lambda t} \tag{4.11}$$

This formula can also be derived by the following considerations: If an original population of X_0 items is continuously decaying or growing so that there are X items at time t, the change of population in one interval dt is dX/dt. Divided by the total population X at t, this gives the positive or negative rate at which the population changes at time t:

$$\pm \lambda = \frac{dX/dt}{X} = \frac{dX}{X} \frac{1}{dt}$$

Rearrangement and integration gives

$$\pm \int_0^t \lambda \, dt = \ln X - \ln C = \ln \frac{X}{C}$$

For $t = 0$, $X = X_0$ and therefore $C = X_0$. Then

$$\frac{X}{X_0} = e^{\pm \int_0^t \lambda dt} = \exp \left[\pm \int_0^t \lambda \, dt \right]$$

and if λ is constant, which is the rate of the population's growth or decay,

$$\frac{X}{X_0} = e^{\pm \lambda t} = \exp \left[\pm \lambda t \right]$$

When λ is positive, the population grows exponentially; when λ is negative, it decays exponentially. Now by definition, if decay is involved, X/X_0 is the probability of survival or reliability $R(t)$, because X is the number surviving by time t out of an original population of X_0. Therefore,

$$R(t) = \exp \left[-\lambda t \right]$$

is the reliability when $\lambda = $ constant, and

$$R(t) = \exp \left[- \int_0^t \lambda \, dt \right]$$

is the general case of reliability when λ, the failure rate, is variable.

We have gone into quite an elaborate derivation of this formula because this is a good opportunity to show how the derivation can help to measure the failure rate of a component population.

Equation (4.8) supplies the method. If λ is constant, the product $(1/N_s)(dN_f/dt)$ must also be constant throughout a test. That means that $1/N_s$ and dN_f/dt must either decrease at the same rate or must be held constant through the entire test. A simple way to measure a constant failure rate is to keep the number of components in the test constant by immediately replacing the failed components with good ones. The number of alive components N_s is then equal to N_0 throughout the test, i.e., equal to the original population because we keep it at N_0 by replacements. Therefore, $1/N_s = 1/N_0$ is constant, and dN_f/dt in this test must also be constant if the failure rate is to be constant. But dN_f/dt will be constant only if the total number of failed components N_f counted from the beginning of the test increases linearly with time. If N_f components have failed in time t at a constant rate, the number of components failing per unit time becomes N_f/t, and in this test we can substitute N_f/t for dN_f/dt and $1/N_0$ for $1/N_s$. Therefore:

$$\lambda = \frac{1}{N_s}\frac{dN_f}{dt} = \frac{1}{N_0}\frac{N_f}{t} \qquad (4.12)$$

Thus, we need to count only the number of failures N_f and the straight hours of operation t. The constant failure rate is then the number of failures divided by the product of test time t and the number of components in test which is kept continuously at N_0. This product N_0t is the number of unit-hours accumulated during the test. Of course, this procedure for determining the failure rate can be applied only if λ is constant.

If only one equipment ($N_0 = 1$) is tested but is repairable so that the test can continue after each failure, the failure rate becomes $\lambda = N_f/t$ where the unit-hours t amount to the straight test time.

In the case of several equipments or components, the denominator is N_0t, which is the sum of the operating times of all specimens under test. It is measured in unit-hours. From (4.12) we also see that the dimension of the failure rate is: number of failures per one unit-hour, or simply "per hour," when t is measured in hours. If time is replaced by the number of cycles, the failure rate can then be expressed as the number of failures per cycle.

In Equation (4.6) we introduced the term dR/dt, the meaning of which will now be discussed. Referring to Figure 3.1 which graphically illustrates the reliability function, we know that dR/dt represents the slope of the R curve at any point t. This slope is always negative and decreases from an initial value of $-1/m$ at $t = 0$ to zero at $t =$ infinity.

The negative character of the slope is expressed by Equation (4.6),

$$\frac{dR}{dt} = -\frac{1}{N_0}\frac{dN_f}{dt}$$

In this equation dN_f/dt is the frequency at which failures occur at any time during a nonreplacement test. When dN_f/dt is plotted as a graph against t, we obtain the time distribution of the failures of all the original N_0 components. And when we plot $(1/N_0)(dN_f/dt)$ as a graph, we have the distribution of failures in time on a per component basis, or the failure frequency curve per component. Therefore the graph $(1/N_0)(dN_f/dt)$ will be of the same character as the graph dN_f/dt, only all ordinates are divided by the constant number N_0. It is thus a unit frequency curve which is called the *failure density function $f(t)$*, or simply, distribution:

$$f(t) = \frac{1}{N_0}\frac{dN_f}{dt} = -\frac{dR}{dt} \tag{4.13}$$

The total area under this curve equals unity, or 100 per cent. The failure rate in Equation (4.9) can therefore also be written as

$$\lambda = -\frac{1}{R(t)}\frac{dR}{dt} = \frac{f(t)}{R(t)} \tag{4.14}$$

which means the failure rate at any time t equals the $f(t)$ value divided by the reliability, both taken at the time t. Equation (4.14) again applies to all possible distributions and reliabilities, whether or not they are exponential.

In the special case when λ is constant, the distribution is

$$f(t) = -\frac{dR}{dt} = -\frac{d}{dt}e^{-\lambda t} = +\lambda e^{-\lambda t} \tag{4.15}$$

The same results can be obtained when Equation (4.4) is used for deriving the failure density curve. We then have

$$f(t) = \frac{1}{N_0}\frac{dN_f}{dt} = \frac{dQ}{dt} \tag{4.16}$$

because $Q(t) = N_f/N_0$ by definition. By integration of (4.16) we obtain

$$Q(t) = \int_0^t f(t)\,dt \tag{4.17}$$

which means that the probability of failure Q at time t is equivalent to the area under the density curve taken from $t = 0$ to t. This area increases for longer operating times t and therefore the probability of failure also increases with t. Thus, $Q(t)$ is the cumulative probability of failure function. Conversely, the probability of survival, or reliability, must decrease for longer operating times t. From $R = 1 - Q$ we get

$$R(t) = 1 - \int_0^t f(t)\, dt$$

but because the area under the density curve is always unity, i.e.,

$$\int_0^\infty f(t)\, dt = 1$$

we can write

$$R(t) = \int_0^\infty f(t)\, dt - \int_0^t f(t)\, dt = \int_t^\infty f(t)\, dt \qquad (4.18)$$

which means that the probability of survival decreases as the remaining

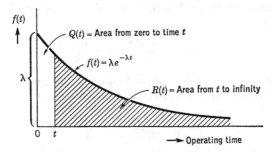

Fig. 4.1. The exponential density function.

area under the density curve decreases. This is shown in Figure 4.1, the graph of the density function for the exponential case. The total area under this curve is always unity:

$$A = \int_0^\infty f(t)\, dt = \int_0^\infty \lambda e^{-\lambda t}\, dt = -\left[e^{-\lambda t} \right]_0^\infty = 1$$

This also applies to the general case when

$$f(t) = -\frac{dR}{dt} = \frac{dQ}{dt}$$

because

$$A = -\int_0^\infty \frac{dR}{dt}\, dt = -\int_0^\infty dR = -[R(\infty) - R(0)] = 1$$

Also the identity

$$A = \int_0^\infty f(t)\, dt = \int_0^t f(t)\, dt + \int_t^\infty f(t)\, dt = Q(t) + R(t) = 1$$

always holds, regardless of what form the distribution takes.

The failure rate can be written also in terms of Q:

$$\lambda = \frac{f(t)}{R(t)} = \frac{f(t)}{1 - Q(t)} = \frac{1}{1 - Q(t)}\frac{dQ}{dt} = +\frac{1}{R(t)}\frac{dQ}{dt} \qquad (4.19)$$

The important point we have made here is that the failure rate is always equal to the ratio of density to reliability. In the exponential case this ratio is constant. However, in the case of nonexponential distributions, the ratio changes with time and therefore the failure rate is then a function of time.

In some cases of military applications, the probability of survival for a short period from time t_1 to time t_2 within the operating time t or within the mission time may be of interest. If a mission of length t is contemplated, $R(t)$ gives the reliability of the entire mission. But what is the probability that the system will function exactly between t_1 and t_2 after taking off at $t = 0$? From the density function the *a priori* probability of failure between t_1 and t_2 is $Q(t_2) - Q(t_1)$ because

$$Q_{t_2-t_1} = \int_{t_1}^{t_2} f(t)\ dt = \left[Q(t) \right]_{t_1}^{t_2} = Q(t_2) - Q(t_1)$$

The probability that the system will not fail in the period from t_1 to t_2 is then:

$$P_{t_2-t_1} = 1 - Q(t_2) + Q(t_1) = 1 - [R(t_1) - R(t_2)] \qquad (4.20)$$

However, this is not the system's reliability of functioning from t_1 to t_2, because the system may fail before t_1, that is, in the period from $t = 0$ to t_1. $P_{t_2-t_1}$ is merely the *a priori* probability that the system will not fail in the specified interval, and this is not identical with the probability that the system will operate in that interval, which is the system's reliability. Stated in terms of probability, system reliability is the probability that the system will survive up to t_2, which is the same as the probability that it will not fail from $t = 0$ to t_1 and also not fail from t_1 to t_2. Therefore,

$$R_{t_2-t_1} = R(t_1) \times R(t_2 - t_1) = R(t_2) \qquad (4.21)$$

Such computations may be of interest in connection with radar guidance for bombing, and similar applications. For instance, if during the first part of a mission, from $t = 0$ to t_1 the system has a failure rate of λ_1, but over the critical target area it has a failure rate λ_2 if it operates at a higher stress level, the system reliability for the critical interval $t_2 - t_1$ is

$$R_{t_2-t_1} = e^{-\lambda_1 t_1} \times e^{-\lambda_2(t_2-t_1)} \qquad (4.22)$$

If $\lambda_1 = 0$, we obtain

$$R_{t_2-t_1} = e^{-\lambda_2(t_2-t_1)} \qquad (4.23)$$

And if $\lambda_1 = \lambda_2 = \lambda$,

$$R_{t_2-t_1} = e^{-\lambda t_2} \qquad (4.24)$$

As seen above, we obtain Equation (4.21) from (4.20) by subtracting $Q(t_1)$. Therefore, quite generally, the probability of surviving an interval $(t_2 - t_1)$ can be obtained from the *a priori* probability of not failing in that interval minus the probability of failing prior to that interval.

Chapter 5

USEFUL LIFE OF COMPONENTS

W E HAVE PREVIOUSLY STATED that when failures are distributed at random intervals in the time domain, the failure rate is constant. "Randomness" implies a certain regularity. If we take a large sample of components and operate them under constant conditions and replace the components as they fail, then approximately the same number of failures will occur in sufficiently long periods of equal length. However, we must at first require that the quality of the components be approximately equal at the beginning of the test and that they not deteriorate during the test. This amounts to the requirement during the test that no component reach a wearout condition or—which amounts to the same— its strength must not deteriorate. The failures which occur are therefore not failures caused by component deterioration but only chance failures, sometimes also called *catastrophic* failures. They are characterized by a sudden breakdown, without preceding deterioration symptoms.

The physical mechanism of such failures is a sudden accumulation of stresses acting on and in the component. It is therefore these sudden stress accumulations which occur at random, and the randomness of the occurrence of chance failures is only a directly observable consequence. If the stress accumulations were to cluster around some preferred time, their occurrence would no longer be completely random. Thus, randomness implies that events occur at unpredictable moments and irregularly, but in sufficiently long periods of equal length approximately the same number of events occur.

When we expose a component population to operation in an environment wherein stress accumulations that cause component failures occur at random, then the component failures themselves also will occur at random. As long as the population is kept constant by replacements, approximately the same number of failures will occur in periods of equal length. But, if we do not replace the failed components, the population will decay exponentially and the number of failures in periods of equal length will also decrease exponentially. This is because there are always less and less components alive which can fail. The mathematical proof of this phenomenon was given in the preceding chapter where we derived the exponential law from the basic notions of probability.

Let us now plot the curve of the failure rate against the lifetime T of a very large sample of a homogeneous component population. The

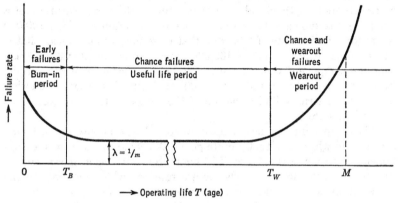

Fig. 5.1. Component failure rate as a function of age.

resulting failure rate graph is shown in Figure 5.1. At the time $T = 0$ we place in operation a very large number of new components of one kind. This population will initially exhibit a high failure rate if it contains some proportion of substandard, weak specimens. As these weak components fail one by one, the failure rate decreases comparatively rapidly during the so-called "burn-in" or "debugging" period, and stabilizes to an approximately constant value at the time T_B, when the weak components have died out. The component population, after having been burned in or debugged, reaches its lowest failure rate level which is approximately constant. This period of life is called the "useful life" period, because it is in this period that the components can be utilized to the greatest advantage, and it is in this period that the exponential law is a good approximation. When the components reach the life T_w, wearout begins to make itself noticeable. From this time on, the failure rate increases rather rapidly. If up to the time T_w only a small percentage of the component population has failed, of the many components which survived up to the time T_w, about one-half will fail in the time period from T_w to M. The time M is the mean wearout life of the population. We call it simply *mean life*, distinguished from the *mean time between failures*, $m = 1/\lambda$ in the useful life period.

The mean life M of components may lie between a few hours and several tens of thousands of hours. There are, of course, exceptions in both directions. The time T_w, up to which the failure rate can be considered constant, is naturally shorter than the mean life M. We shall consider the relationship between T_w and M when discussing the normal distribution of wearout.

The mean time between failures of components, $m = 1/\lambda$ is usually much larger than the mean wearout life M. If the chance failure rate is very small in the useful life period, the mean time between failures can reach hundreds of thousands and even millions of hours. Naturally, if a component is known to have a mean time between failures of say 100,000 hours (or a failure rate of 0.00001), that certainly does not mean that it can be used in operation for 100,000 hours. From Chapter 3 we know that the component would have only a 36.8 per cent chance to survive its mean time between failures of 100,000 hours. But there is another factor which strictly limits the life of the component. It is wearout, as shown in Figure 5.1.

The component may have a mean wearout life M of only 10,000 hours, a mean life which is actually quite good. We could operate such a component up to perhaps a T_w of 6000 hours, but not longer, if we want to prevent a steep increase of the failure rate or—which amounts to the same—if we want to prevent the component from failing in operation. Therefore, the very low failure rate which gives a mean time between failures of 100,000 hours exists only in the first 6000 hours of the component's life. If the component does not fail up to the 6000 hours, which it probably will not do, we have to replace it by a new and equally good component if we want to continue operation at the high reliability level of $m = 100,000$ hours.

What, then, is the meaning of the mean time between failures m? It simply tells us how reliable the component is in its useful life period, and such information is of utmost importance. A component with a mean time between failures of 100,000 hours will have a reliability of 0.9999 or 99.99 per cent for any 10-hour operating period. Further, if we operate 100,000 components of this quality for 1 hour, we would expect only one to fail. Equally, would we expect only one failure if we operate 10,000 components under the same conditions for 10 hours, or 1000 components for 100 hours, or 100 components for 1000 hours. These components can, under circumstances, yield reliable systems as long as we do not allow any of them to remain in service for more than T_w hours. We must preventively replace each component before it gets a chance to fail from wearout.

For instance, if a system contains one hundred components of this quality, chance failures will occur occasionally, but wearout failures, which can cause much more trouble, will not occur as long as we abide by the preventive replacement rule. With 100 of these components in a system we would expect on the average only one failure in 1000 hours, or 6 failures in 6000 hours. These would be only chance failures, not wearout failures. We prevent the wearout failures by timely, preventive replacement.

On the other hand, chance failures cannot be prevented by any replacement policy, because of the constant failure rate of the components within their useful life. If we tried to replace good nonfailed components after, say 1000 hours, we would improve absolutely nothing. We would more likely do harm, as some of the components used for replacement may not have been properly burned in, and the presence of such components could only increase the failure rate. Therefore, the very best policy in the useful life period of components is to replace them only as they fail. However, we must stress again that no component must be allowed to remain in service beyond its wearout replacement time T_w. Otherwise, the component probability of failure increases tremendously and the system probability of failure increases even more. The reliability of systems where no preventive replacement of components at the proper time is practiced can drop to ridiculously low values.

The golden rule of reliability is therefore: Replace components as they fail within the useful life of the components, and replace each component preventively, even if it has not failed, not later than when it has reached the end of its useful life. Figure 5.1 also shows that the failure rate may be quite high initially when new components, or new systems, are placed into operation. This high failure rate is due to so-called "early failures." Substandard components usually fail in the first few hours of operation. However, the debugging or burn-in period may last up to several hundred hours if the number of substandard components is large. In this period many failures may occur, but when we replace the weak components by good ones or repair other initial faults, the failure rate soon drops to an essentially constant value. If it is undesirable to have these early failures causing trouble in the early life of an equipment, it becomes necessary to burn in the components for the time T_B before using them for equipment assembly. Such burn-in weeds out the substandard components, and the equipment then experiences much less component failures from the very beginning of its operation. The burn-in procedure is an absolute must for missile, rocket, and space systems in which no component replacements are possible once the vehicle takes off and where the failure of any single component can cause the loss of the system.

Component burn-in before assembly followed by a debugging procedure of the system is therefore another golden rule of reliability. It applies especially to missiles and similar systems, in view of their high replacement cost. Unfortunately, with a few exceptions, there is still a lot of trespassing against this rule by the suppliers of components.

Chapter 6

WEAROUT AND RELIABILITY

HIGH SYSTEM RELIABILITIES can be achieved for extended periods of operation with proper debugging procedures to eliminate early failures and with strict methods of component preventive replacement so that the components do not get a chance to fail because of wearout. Component replacement is essential if reliable operation is required beyond the component's wearout time T_w in Figure 5.1.

Such replacements restore the equipment or system to an operational condition of low probability of failure. Thus, when good preventive maintenance is practiced, reliable system operation for very long periods becomes possible. A system which is regularly overhauled at appropriate overhaul times almost never ages. When components are not allowed to wear out in a system, the system will not fail because of wearout. Although it may still exhibit some chance failure rate, it will, in general, be very reliable.

High reliabilities imply low chance failure rates and therefore long mean times between failures. As was shown in Chapter 3, the mean time between failures is a very convenient parameter to express and calculate reliability in the useful life period. When the mean time between failures of an equipment is known, its reliability for an operating period of a given length can be calculated immediately by means of the exponential formula. The mean time between failures is an average time at which failures occur. It is thus an average time for which failure-free operation is expected. Since it is only an average, we must expect that there will be instances when failures occur in much shorter periods of operation than the mean time between failures. Equally, there will be instances when failures occur only after much longer operating times. It is therefore not admissible to conclude from a known mean time between failures of

m hours that the equipment or system will operate without failure for these m hours.

In contrast to the normal or Gaussian failure distribution, which usually approximates wearout phenomena quite well, and where about one-half of the failures occur before the mean life and one-half occur later, in the exponential distribution about 63 per cent of the failures occur in operating times shorter than the mean time between failures, and about 37 per cent of the failures occur only after operating times longer than the mean time between failures.

In the exponential distribution the frequency of failures is higher towards the shorter operating times. Therefore, reliable operation can be achieved only for operating times much shorter than the mean time between failures. Only for operating times that are short compared with m do low probabilities of failure exist, and therefore high probabilities of failure-free operation.

On the other hand, in the case of the normal distribution the failures cluster around the mean life M. Failure-free operation can therefore often be achieved up to a component age relatively close to the mean life of the components, according to how widely the normal curve is spread in a given case.

In order to show the basically different character of the exponential and normal distributions, their failure density functions are compared in Figure 6.1. These curves illustrate the density at which component failures of the original population occur in the time domain when failed components are not replaced and all components are allowed to die out. The density function of the exponential distribution is given (according to Equation (4.15)) by $f(t) = \lambda e^{-\lambda t}$, where t is the time to failure counted from any arbitrary time origin $t = 0$ at which the component was still alive. Therefore the exponential distribution is independent of component age, but, of course, only as long as the component strength has not deteriorated, or as long as the failure rate λ is constant. When strength deterioration sets in, the failure rate begins to rise, and wearout, which usually follows a normal distribution, is superimposed on the exponential failure frequency.

The normal density function is given by

$$f(T) = \frac{1}{\sigma \sqrt{2\pi}} e^{-(T-M)^2/2\sigma^2} \qquad (6.1)$$

where T is the component age. Thus, the normal distribution depends upon age, whereas the exponential does not! We shall prove this statement later.

Figure 6.1 shows that an exponentially behaving population of components, when placed in a test in which the failed components are not replaced, suffers its greatest losses in the test period before m, whereas a normal population of components suffers its greatest losses around the time M, its mean life or mean age.

From the density curves we obtain the probability of failure by integration as the area under these curves. These probabilities of failure are

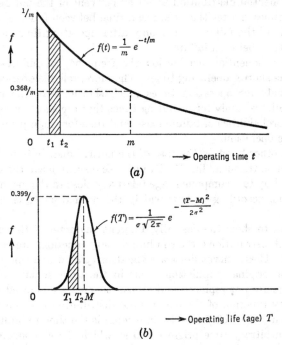

Fig. 6.1. The exponential and normal density functions. (a) Exponential density function. (b) Normal density function.

shown for different time intervals $t_2 - t_1$ and $T_2 - T_1$ by the shaded areas. They amount to

$$P_{t_2-t_1} = \int_{t_1}^{t_2} f(t)\, dt \quad \text{and} \quad P_{T_2-T_1} = \int_{T_1}^{T_2} f(T)\, dT \tag{6.2}$$

We call P the *a priori probability of failure for a given time interval from t_1 to t_2 or T_1 to T_2.*

The total area under the density curve equals unity by definition. Stated verbally, the probability of failure for the infinite time interval under the whole density curve is 100 per cent, or certainty. For the exponential case the result reads

$$\int_0^\infty \lambda e^{-\lambda t}\, dt = -\frac{\lambda}{\lambda}\left[e^{-\lambda t} \right]_0^\infty = 1 \tag{6.3}$$

When we take the integral from $t = 0$ to an arbitrary time t, we obtain the *cumulative probability of failure* for the period from zero time to time t, which we call unreliability $Q(t)$:

$$Q(t) = \int_0^t f(t)\, dt \qquad (6.4)$$

The equation plotted in graph form for various values of t gives the

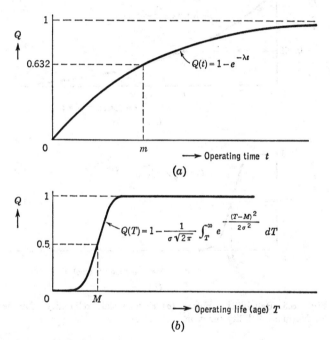

Fig. 6.2. Probability of failure curves. (*a*) Exponential. (*b*) Normal.

cumulative probability distribution curve, or the unreliability curve. In the exponential case we take the values of $\int_0^t f(t)\, dt$, and use these values as ordinates and the times t as abscissae. The equation of this curve is

$$Q(t) = \int_0^t \lambda e^{-\lambda t}\, dt = -\left[e^{-\lambda t}\right]_0^t = 1 - e^{-\lambda t} \qquad (6.5)$$

When plotting the cumulative probability of failure curves or Q curves for both the exponential and normal distribution, we obtain the graphs shown in Figure 6.2. As to the probability of survival R for the same time period, this is $R = 1 - Q$, and therefore the R curves or probability of survival curves of the two distributions look as shown in Figure 6.3. These curves are also called the *survival characteristics*. In the exponential case the probability of survival $R(t)$ decreases with t relative to m (for the first hours of operation) much faster than $R(T)$ decreases in the normal case with T relative to M.

The graphs shown are not drawn to any actual scales. They are used only to illustrate the dissimilarity of behavior in the time domain of the

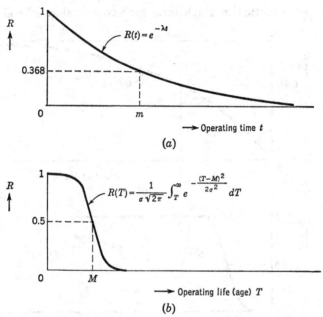

Fig. 6.3. Reliability curves. (a) Exponential reliability function. (b) Normal reliability function.

two characteristic types of failures and their probabilities—the chance failures which are exponentially distributed, and the wearout failures which are approximately normally distributed. In most component cases the values of M are comparatively much shorter than those of m.

The value of the mean time between failures m of very good components operated in an ideal environment may become even millions of hours—by human measure approaching infinity or absolute reliability. But the M value of components, i.e., their mean life, is usually limited to a few thousand hours, or in some exceptional cases to a few tens of thousands. Therefore, even if a component could operate with almost absolute reliability within its useful life period, this high reliability will persist only for a comparatively short lifetime T_w. If not chance, then wearout will definitely cause it to fail if it is not pulled out from service before this happens.

Of course, from an economic viewpoint we might ask what advantage is gained by removing a component from service before it has failed and inserting a new one. Would it not be more economical to let it operate until it wears out and fails? Then surely less spares would be needed in the long run.

Such an approach can be very deceptive. A preventive replacement can be made when the system in which the component operates is temporarily not in service, such as during a regular overhaul or during any scheduled break in the system's operation. Therefore, a preventive replacement normally does not involve any losses which would otherwise be incurred by having to stop the system during its regular operational time when some component fails. Obviously, component failure during operation may result in the complete loss of the infinitely more costly system and, in aircraft operations, perhaps even loss of lives. Therefore, whenever components are involved which can cause system breakdowns, complete loss of system, or loss of lives, a preventive replacement of such components is a better economic proposition than waiting until they fail in operation. Preventive replacement is a must for reliable operation of long-life systems.

Let us now look at how reliability is affected if component wearout is not systematically prevented by timely replacement. In most cases wearout displays typical characteristics of a normal or Gaussian distribution. The formula of the Gaussian normal density curve is given by

$$f(T) = \frac{1}{\sigma \sqrt{2\pi}} \, e^{-(T-M)^2/2\sigma^2} \qquad (6.6)$$

where M is the mean wearout life, T is the age or accumulated operating time since new, and σ is the standard deviation of the lifetimes from the mean M, defined as

$$\sigma = \sqrt{\frac{\Sigma \, (T - M)^2}{N}} \qquad (6.7)$$

The value N in σ is the number of events (failures or, which amounts to the same, lifetimes T) over which the summation $\Sigma \, (T - M)^2$ is made. The total area under the density curve $f(T)$ is again unity, or 100 per cent, as shown in Figure 6.4. Any partial area under this curve from T_1 to T_2 represents the percentage of the original N components which fail in the time interval $T_2 - T_1$. The partial area also represents the *a priori* probability of any individual component of the original population to fail in the interval $T_2 - T_1$, when put into operation at the time $T = 0$ when new. Care must be exercised in the interpretation of *a priori* probabilities. For instance, the *a priori* probability of a component to fail in the time interval

$M - 3\sigma$ to $M - 2\sigma$ is 2.14 per cent; in the time from $M + 2\sigma$ to $M + 3\sigma$ it is again 2.14 per cent.

Thus, judged at the time $T = 0$ when the component is new, it has the same probability of 2.14 per cent of failing in the time interval $M - 3\sigma$ to $M - 2\sigma$ as in the time interval $M + 2\sigma$ to $M + 3\sigma$. It has the same chance of being a worse-than-average component or a better-than-average component. However, these two equal 2.14 per cent probabilities do not have the same meaning from the reliability standpoint, or from the standpoint of the component's probability of survival.

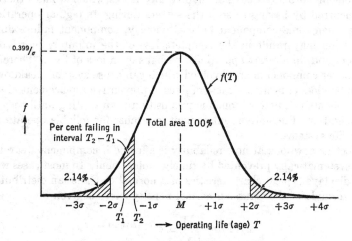

Fig. 6.4. The *a priori* probability of failure.

In the first case, the component has a probability to survive from the time it was new ($T = 0$) to the time $T = M - 3\sigma$ (i.e., the time when the interval $M - 3\sigma$ to $M - 2\sigma$ begins) of 99.865 per cent. But in the second case it has a probability of surviving from $T = 0$ to $T = M + 2\sigma$ of only 2.275 per cent. Thus, the 2.14 per cent probability of failure in the two time intervals has to be weighed against the probability of surviving up to the beginning of the respective time interval. The statement that the component has a probability of failure of only 2.14 per cent in the time interval from $M + 2\sigma$ to $M + 3\sigma$ has no reliability meaning unless we add that its probability of failing before the time $M + 2\sigma$ is $100 - 2.275 = 97.725$ per cent. It is therefore doubtful whether the component will ever reach the age $M + 2\sigma$. The 2.14 per cent probability of failure is only an *a priori* probability which has to be added to the probability of failure from $T = 0$ to the time when the interval begins. Thus, in the second case, we have the total probability of failure of $97.725 + 2.14 = 99.865$ per cent. This is the cumulative probability of failure of the component from $T = 0$ to $T = M + 3\sigma$. Therefore, the component has a chance of surviving from $T = 0$ up to the time $M + 3\sigma$ of only $100 - 99.865 = 0.135$ per cent, whereas in the first case it has a cumulative probability of failure before $M - 2\sigma$ of $0.135 + 2.14 = 2.275$ per cent, and therefore a probability of surviving from $T = 0$ up to the time $M - 2\sigma$ of $100 - 2.275 = 97.725$ per cent. For a realistic evaluation of the chance of a component to survive up to a given age T, the cumulative probabilities are required. They can be found from normal probability tables.*

But in certain cases we still may want to know the probability of a component's failure or survival in a given time interval of the normal curve, if the component survives up to the beginning of that time interval. Here the condition "if it survives" is introduced, and this creates the case of a *conditional probability*.

We can calculate the conditional probabilities for the time intervals $M - 3\sigma$ to $M - 2\sigma$, and $M + 2\sigma$ to $M + 3\sigma$ by asking what the probability of failure or of survival is in the respective time interval if, or given that, the component survives up to the beginning of that time interval. Using the theorem on conditional probabilities, the probability of failure in the time interval $T_2 - T_1$, given that the component has survived up to the time T_1, is the *a priori* probability of failure in the time interval $T_2 - T_1$ divided by the cumulative probability of surviving from $T = 0$ to T_1:

$$F_{T_2 - T_1} = \frac{\int_{T_1}^{T_2} f(T)\, dT}{R(T_1)} \tag{6.8}$$

This value is the *a posteriori* probability of failure in the time interval T_1 to T_2. In the first case of our example then we have

$$F_{M-3\sigma \text{ to } M-2\sigma} = \frac{0.0214}{0.99865} = 0.02143$$

and a probability of survival in that interval of

$$R_{M-3\sigma \text{ to } M-2\sigma} = 1 - F = 0.97857$$

In the second case we have

$$F_{M+2\sigma \text{ to } M+3\sigma} = \frac{0.0214}{0.02275} = 0.941$$

or a probability of survival in that interval of

$$R_{M+2\sigma \text{ to } M+3\sigma} = 1 - F = 0.059$$

* For instance, *Tables of Normal Probability Functions*, National Bureau of Standards, Applied Mathematics Series No. 23, U.S. Government Printing Office, Washington, D.C., 1953.

It would obviously make no sense to say that the component has a probability of survival of $1 - 0.0214 = 0.9786$ in the interval $M + 2\sigma$ to $M + 3\sigma$ because the *a priori* probability of failure P is 0.0214 in that interval. But we can predict that at the time $T = 0$ there is a 97.86 per cent probability that the component will not fail in the interval $M + 2\sigma$ to $M + 3\sigma$, although it will almost certainly fail earlier.

As the reader has seen, we have used three different symbols for probabilities of failure in the preceding paragraphs:

1. $P_{T_2-T_1}$ as the *a priori* probability of failure in the time interval from T_1 to T_2,
2. $F_{T_2-T_1}$ as the *a posteriori* or conditional probability of failure in the same time interval $T_2 - T_1$,
3. $Q(T)$ as the cumulative probability of failure from $T = 0$ to the time T.

The three probabilities are connected by the following formulas derived from (6.2), (6.4), and (6.8):

$$P_{T_2-T_1} = \int_{T_1}^{T_2} f(T)\, dT = Q(T_2) - Q(T_1) \tag{6.9}$$

$$F_{T_2-T_1} = \frac{P_{T_2-T_1}}{R(T_1)} = \frac{Q(T_2) - Q(T_1)}{R(T_1)} \tag{6.10}$$

$$Q(T) = \int_0^T f(T)\, dT = 1 - \int_T^\infty f(T)\, dT = 1 - R(T) \tag{6.11}$$

We designate $F_{T_2-T_1}$ and $Q(T)$ by the word *unreliability*. $F_{T_2-T_1}$ is the conditional unreliability for a time interval $T_2 - T_1$ given that no failure occurs up to T_1, and $Q(T)$ is the cumulative unreliability for the time interval from $T = 0$ to T.

The difference between the *a priori* and *a posteriori* probabilities becomes much clearer when we investigate it from the viewpoint of the basic definition of probability.

Let the normal density curve represent the wearout failure frequency of a population of 10,000 components new at the time $T = 0$. One then expects 214 components, or 2.14 per cent of 10,000, to fail in the interval from $M - 3\sigma$ to $M - 2\sigma$ and 214 components (again, 2.14 per cent of 10,000) to fail in the interval from $M + 2\sigma$ to $M + 3\sigma$ also. These 2.14 percentages are *a priori* probabilities of failure in the sense that they refer to the entire population of 10,000 components while none of them has yet failed. However, as the experiment progresses, components begin to fail and each subsequent period begins with less and less components alive. Now the probability of failure or unreliability in a given interval is defined as the number failing in that interval divided by the total number of components living at the beginning of this interval. This, in our case, amounts to the conditional unreliability. In the interval from $M - 3\sigma$ to $M - 2\sigma$, 214 components will fail out of 9986 alive at $M - 3\sigma$; therefore, the conditional probability of failure is $^{214}/_{9986} = 0.02143$ or 2.143 per cent. In the interval from $M + 2\sigma$ to $M + 3\sigma$, again 214 components will fail. However, only 228 are alive at $M + 2\sigma$, so the conditional probability of failure is $^{214}/_{228} = 0.941$ or 94.1 per cent. Obviously, it makes a tremendous difference whether 214 components fail out of 9986 or out of

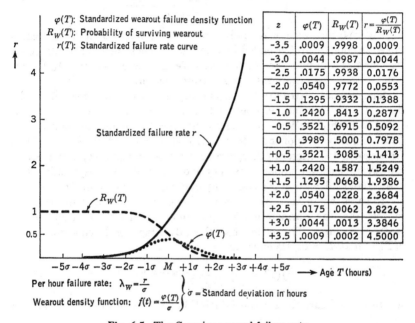

z	$\varphi(T)$	$R_W(T)$	$r = \dfrac{\varphi(T)}{R_W(T)}$
-3.5	.0009	.9998	0.0009
-3.0	.0044	.9987	0.0044
-2.5	.0175	.9938	0.0176
-2.0	.0540	.9772	0.0553
-1.5	.1295	.9332	0.1388
-1.0	.2420	.8413	0.2877
-0.5	.3521	.6915	0.5092
0	.3989	.5000	0.7978
+0.5	.3521	.3085	1.1413
+1.0	.2420	.1587	1.5249
+1.5	.1295	.0668	1.9386
+2.0	.0540	.0228	2.3684
+2.5	.0175	.0062	2.8226
+3.0	.0044	.0013	3.3846
+3.5	.0009	.0002	4.5000

$\varphi(T)$: Standardized wearout failure density function
$R_W(T)$: Probability of surviving wearout
$r(T)$: Standardized failure rate curve

Per hour failure rate: $\lambda_W = \dfrac{r}{\sigma}$
Wearout density function: $f(t) = \dfrac{\varphi(T)}{\sigma}$ $\Big\}$ σ = Standard deviation in hours

Fig. 6.5. The Gaussian normal failure rate.

228 in a time interval of the same length; the failure rate of the components has phenomenally increased by the time $M + 2\sigma$, as compared to the failure rate at the time $M - 3\sigma$. Such a dramatic failure rate increase is characteristic of wearout.

The increase of the failure rate can be shown in a failure rate graph for the normal distribution of failures. Equation (4.14) defined the failure rate as $\lambda = f(t)/R(t)$ or $f(T)/R(T)$. When we evaluate this ratio from the instantaneous values of the $f(T)$ curve (dotted) and the $R(T)$ curve (dashed) of the normal distribution, we obtain the failure rate curve shown by the plain line in Figure 6.5. To standardize this curve, the standardized failure density function in units of the standard deviation is used:

$$\varphi(T) = \sigma f(T)$$

The values of $\varphi(T)$ together with $R_w(T)$ can be obtained directly from normal tables. In this way we obtain the standardized failure rate curve as

$$r(T) = \frac{\varphi(T)}{R_w(T)}$$

The per-hour wearout failure rate is then

$$\lambda_w = \frac{f(T)}{R_w(T)} = \frac{\varphi(T)}{\sigma R_w(T)} = \frac{r(T)}{\sigma}$$

Figure 6.5 shows how the failure rate of a component rapidly increases once the component has ended its useful life. When the life T is measured in hours, λ_w will be in per hour units. We use the designation λ_w to emphasize that this failure rate results from wearout.

The normal distribution has the difficulty that the area under the density curve becomes 100 per cent only if the curve is extended to infinity in both directions. This is not possible for time-dependent events because a new component enters service at $T = 0$ and not at $-\infty$. The Gaussian distribution can therefore be regarded only as an approximation. However, this approximation is very good in most cases, especially when the standard deviation is small compared with the mean life M.

In cases where $M < 3\sigma$, one would have to consider a transformation of the Gaussian normal to the logarithmic normal distribution of the wearout failures. The latter has the advantage of having the value $f(T) = 0$ at $T = 0$. Its equation is

$$f(T) = \frac{1}{\sigma T \sqrt{2\pi}} e^{-(\log T - M)^2/2\sigma^2} \tag{6.12}$$

where σ and M are the standard deviation and the mean of the logarithm of T.*

Its cumulative distribution, or probability of failure from $T = 0$ to T is, as before,

$$Q(T) = \int_0^T f(T)\, dT$$

In the case of the Gaussian normal distribution, the lower time limit of the integral would have to be $-\infty$, which makes no sense. This difficulty with the Gaussian curve is overcome by integrating from T to $+\infty$ and taking the value of the integral away from 1. This operation is correct because the area under the density curve equals unity:

$$Q(T) = 1 - \int_T^{+\infty} f(T)\, dT \tag{6.13}$$

---

* When working with the log normal distribution, change values of T to $\log T$, and treat $\log T$ as normally distributed. From normal tables obtain the answers and transform back to T.

and the second term on the right is then the cumulative reliability from $T = 0$ to T:

$$R(T) = \int_T^{+\infty} f(T)\, dT \qquad (6.14)$$

In further discussions the Gaussian normal distribution will be used in preference to the logarithmic normal distribution. It is more familiar to most engineers and is easy to evaluate.

Having reviewed the normal distribution approximating wearout, we can now determine the age T_w at which a component has to be pulled out of service and replaced to prevent wearout failures in operation. The replacement time T_w has to be chosen so as to keep the cumulative probability of wearout failure $Q(T)$ at a minimum acceptable level, consistent with the reliability requirements of the particular case. $Q(T)$ is defined by Equation (6.13).

To avoid mistakes, we shall use the subscript w for wearout probabilities. For instance, if $T_w = M - 3\sigma$ is chosen as replacement time, the probability of wearout failure for the entire operating life of the component from $T = 0$ to $T = T_w = M - 3\sigma$ amounts to $Q_w = 0.00135$. This value is obtained from the cumulative probability tables of the normal distribution. For a replacement time $T_w = M - 4\sigma$, we find $Q_w = 0.0000317$, and for $T_w = M - 5\sigma$, $Q_w = 0.000000287$, etc. These figures seem to be very low for the useful life of a component. However, when there are thousands or tens of thousands of components in series in a system, the wearout probabilities of the individual components add up and the system's probability of surviving wearout drops rapidly. Therefore, the choice of the replacement time must be very carefully considered.

Let us look at a single-engine airplane. If the engine is replaced or overhauled every $T_w = M - 3\sigma$ hours, its probability of failing from wearout in flight between two overhauls is $Q_w = 0.00135$. We could then expect about one engine wearout failure in flight for about 700 overhauls. This sounds very good. But if we choose to replace at $T_w = M - 2\sigma$, the failure probability jumps to $Q_w = 0.0228$ between two overhauls—or at least one engine wearout failure in flight for 50 overhauls. For 50 single-engine airplanes this would mean an average of one engine failure in flight between every two overhauls of the fleet.

If we now consider an electronic system with 10,000 equal or similar components in series, we may be surprised to discover that replacing the components at $T_w = M - 3\sigma$ would be no good at all. Even $T_w = M - 4\sigma$ is still quite insufficient, because $Q_w = 0.0000317$ and $R_w = 0.9999683$ for each component, and this raised to the power of 10,000 results in a ridiculously low system reliability—the system would almost certainly fail of wearout during operation between each two scheduled overhauls. To keep such a system reasonably immune to wearout failures in operation, the component replacement time would have to be chosen as $T_w = M - 5\sigma$ or even $M - 6\sigma$, according to the system reliability requirements.

As we have seen from the normal curve and from its failure rate curve, the probability of wearout failure increases with age—extremely slowly at first, but as the age approaches the mean wearout life, this increase becomes very rapid.

Within the useful life period there are differences in the conditional failure probabilities F_w for operating intervals of equal lengths. For instance, F_w for the first 10 hours of the component's operational life (from $T = 0$ to $T = 10$) will be orders of magnitude smaller than the F_w for the last 10 hours before T_w (from $T = T_w - 10$ to $T = T_w$). These probabilities can be calculated by the use of Equation (6.8).

If T_w is properly chosen, the probability of surviving wearout (R_w) up to T_w is so close to 100 per cent that the denominator in (6.8) can be taken as equalling unity. Therefore F_w is approximately equal to the partial area under the density curve for the operating interval $T_2 - T_1$, i.e., approximately equal to the *a priori* probability of failure

$$F_{w(T_2-T_1)} \cong P_{T_2-T_1} = Q_w(T_2) - Q_w(T_1) \tag{6.15}$$

This approximation can be used for the normal curve up to about $T = M - 2\sigma$, but not for a later component age.

The increasing probability of failure with component age, so characteristic of wearout situations, is in complete contrast to the probability of chance failures which are distributed exponentially. In the exponential distribution, the probability of failure is independent of the component age and is therefore the same for a given mission time t when the component is new as for a much later mission time of the same t hours of duration, when the component may have already accumulated thousands of hours of operation, if its strength has not suffered any appreciable deterioration.

To prove this, we ask for the probability of failure of an exponential component in a $t = 10$ hour period when the component has already an age T, i.e., when it has already previously operated for T hours. This is an *a posteriori* or conditional probability because it implies that the component survives up to T, and we ask for the probability of failure in the $t = 10$ hours subsequent to T hours, i.e., from T to $T + 10$ hours. We start from the exponential density function $f(t) = \lambda e^{-\lambda t}$, form its integral from T to $T + t$, which is

$$\lambda \int_{T}^{T+t} e^{-\lambda t}\, dt = -\frac{\lambda}{\lambda}\left[e^{-\lambda t}\right]_{T}^{T+t} = e^{-\lambda T} - e^{-\lambda(T+t)} = e^{-\lambda T} - e^{-\lambda T}\cdot e^{-\lambda t}$$

$$\tag{6.16}$$

and which amounts to the *a priori* probability of failure P in the period

T to $T + t$. To obtain the *a posteriori* or conditional probability of failure F which we are seeking, we divide the *a priori* probability of failure by the probability of survival up to T, which is $e^{-\lambda T}$ and which cancels out as follows:

$$F = \frac{e^{-\lambda T} - e^{-\lambda T} \cdot e^{-\lambda t}}{e^{-\lambda T}} = 1 - e^{-\lambda t} \qquad (6.17)$$

But this is also equal to the cumulative probability of failure from $T = 0$ to $T = t$, and therefore, in the exponential case

$$F(t) = Q(t) = 1 - e^{-\lambda t} \qquad (6.18)$$

which proves that the exponential probability of failure in an arbitrary time interval t is independent of the age T of the component. Consequently also the reliability $R = e^{-\lambda t}$ remains the same for an operating interval of length t throughout the useful life of the component. This also means that the chance failure rate

$$\lambda = \frac{f(T)}{R(T)} = \frac{\lambda e^{-\lambda T}}{e^{-\lambda T}} = \lambda$$

is independent of the age T, and remains constant for a given stress level.

If the combined effects of chance and wearout are to be evaluated for a mission time of t hours and when the component has an age of T hours at the beginning of the mission, the combined probability of component failure in t equals the probability of failing of chance and/or wearout in the interval t, at the age of T:

$$Q(t) = Q_c(t) + F_w(t) - Q_c(t) \cdot F_w(t) \qquad (6.19)$$

where the subscripts c and w stand for chance and wearout, where $Q_c(t) = 1 - e^{-\lambda t}$ is independent of the age T, and where, according to (6.8) and (6.14),

$$F_w(t) = \frac{\dfrac{1}{\sigma \sqrt{2\pi}} \displaystyle\int_T^{T+t} e^{-(T-M)^2/2\sigma^2} \, dT}{\dfrac{1}{\sigma \sqrt{2\pi}} \displaystyle\int_T^{\infty} e^{-(T-M)^2/2\sigma^2} \, dT}$$

and depends on the age T of the component.

The component reliability for the mission time t is then

$$R(t) = e^{-\lambda t} - e^{-\lambda t} \cdot F_w(t) = e^{-\lambda t}[1 - F_w(t)] = e^{-\lambda t} \cdot R_w(t) \qquad (6.20)$$

where

$$R_w(t) = 1 - F_w(t) = \frac{\dfrac{1}{\sigma \sqrt{2\pi}} \displaystyle\int_{T+t}^{\infty} e^{-(T-M)^2/2\sigma^2} \, dT}{\dfrac{1}{\sigma \sqrt{2\pi}} \displaystyle\int_T^{\infty} e^{-(T-M)^2/2\sigma^2} \, dT} \qquad (6.21)$$

is the probability that the component will not fail of wearout during the mission of t hours from T to $T + t$. This quotient can easily be determined by dividing the respective areas found from the normal tables. The two areas are the probabilities of surviving wearout up to $T + t$ and up to T. Therefore, the probability that no wearout failure will occur in the interval t is

$$R_w(t) = \frac{R_w(T + t)}{R_w(T)} \tag{6.22}$$

When chance is included, the over-all probability of no failure, or reliability, of the component for the interval t from T to $T + t$ is:

$$R(t) = e^{-\lambda t} \cdot \frac{R_w(T + t)}{R_w(T)} \tag{6.23}$$

When components are combined in series into a system, the system reliability $R_s(t)$ will equal the product of all component reliabilities in that period t:

$$R_s(t) = \exp\left[-\Sigma\,\lambda_i t\right] \cdot \Pi\,\frac{R_{wi}(T_i + t)}{R_{wi}(T_i)}{}^* \tag{6.24}$$

As long as the product $\Pi\,\dfrac{R_{wi}(T_i + t)}{R_{wi}(T_i)}$ is kept close to unity—for instance, 0.999 or higher in the case of an essential aircraft subsystem—the subsystem's reliability can be calculated from the exponential component reliabilities.

In the simple series case we then have:

$$R_s(t) = \exp\left[-\Sigma\,\lambda_i t\right] \tag{6.25}$$

The requirement to hold the wearout product $\Pi\,\dfrac{R_{wi}(T_i + t)}{R_{wi}(T_i)}$ as close to unity as possible also defines the replacement time T_{wi} of each component, where T_{wi} stands for the replacement time of the ith component or, in other words, T_{wi} is the age of the ith component at which it is replaced. For instance, if the wearout product for all components in the system must not drop below 0.999 during any mission time of length t, each component must be replaced by a new one at an age T_{wi} such that

$$\Pi\,\frac{R_{wi}(T_{wi})}{R_{wi}(T_{wi} - t)} \geqq 0.999$$

$R_{wi}(T_{wi})$ denotes the probability that the ith component will not fail because of wearout in the period from the time it was new up to its accumulated operating age T_{wi} at which it is replaced, and $R_{wi}(T_{wi} - t)$ denotes its reliability up to an accumulated operating age $T_{wi} - t$, where t is the system's mission time of length t. Different components will have

---

* The symbol Π stands for *product*, as the symbol Σ stands for *summation;* frequent use will be made of both symbols. The symbol Π must be distinguished from the number π which is also frequently used in the text.

to be pulled out of service and replaced or overhauled at different component ages T_{wi} unless, for economy reasons, a common T_w in terms of the system time is fixed for all components. In such case, some components will be replaced earlier than necessary, but this procedure will improve system reliability, and maintenance cost is usually reduced when many components are replaced at one time rather than separately at different times.

Finally, the over-all reliability of a component as defined by Equation (6.23) can also be expressed in terms of its over-all failure rate by means of Equation (4.10). When we denote by $\lambda = \lambda_c + \lambda_w$ the over-all failure rate as the sum of the chance λ_c and wearout λ_w failure rates, we can write

$$R(t) = e^{-\lambda_c t} \cdot \frac{R_w(T + t)}{R_w(T)} = \exp\left[-\int_T^{T+t} \lambda \, dt\right] \qquad (6.26)$$

The wearout failure rate at any given age of the component can be obtained from tables of the normal distribution as shown in Figure 6.5, where $\lambda_w = r/\sigma$ on a per hour basis, so that it can be added to the chance failure rate λ_c. The exponent of integral form in Equation (6.26) is then the area under the $\lambda = \lambda_c + \lambda_w$ curve taken from T to $T + t$:

$$\int_T^{T+t} \lambda \, dt = \int_T^{T+t} (\lambda_c + \lambda_w) \, dt = \lambda_c t + \int_T^{T+t} \lambda_w \, dt \qquad (6.27)$$

The integral $\int_T^{T+t} \lambda_w \, dt$ for short intervals of t can be approximated by using the arithmetic mean at the values of T and $T + t$, i.e.,

$$\lambda_{wm} = \frac{1}{2} [\lambda_w(T) + \lambda_w(T + t)]$$

which can also be written in terms of the normal frequency function and the normal cumulative probability of survival function,

$$\lambda_{wm} = \frac{1}{2} \left[\frac{f(T)}{R(T)} + \frac{f(T + t)}{R(T + t)}\right]$$

The integral then becomes approximately

$$\int_T^{T+t} \lambda_w \, dt = \lambda_{wm} t \qquad (6.28)$$

and the combined reliability approximately

$$R(t) = \exp[-(\lambda_c + \lambda_{wm})t] \qquad (6.29)$$

Here λ_{wm} is obviously a function of the component's age, T.

As long as λ_{wm} is at least one order of magnitude smaller than the chance failure rate λ_c, reliability is well approximated by $R(t) = e^{-\lambda_c t}$, because the contribution of wearout to the over-all failures will remain below 10 per cent. This procedure can also be used for the determination of the component replacement age, T_W.

Chapter 7

COMBINED EFFECTS OF
CHANCE AND WEAROUT FAILURES

THE COMBINED CUMULATIVE PROBABILITY of a component to survive chance and wearout can be derived from Equation (6.23) for the life period from $T = 0$, when the component is new, to an age T

$$R(T) = e^{-\lambda T} \frac{R_w(0 + T)}{R_w(0)} = e^{-\lambda T} \cdot R_w(T) \qquad (7.1a)$$

because $R_w(0)$ is assumed to be one. In this formula,

$$R_w(T) = \frac{1}{\sigma \sqrt{2\pi}} \int_T^{\infty} e^{-(T-M)^2/2\sigma^2} \, dT \qquad (7.1b)$$

where T = operating age of the component and M is its mean wearout life.

Figure 7.1 shows the reliability curve $R(T)$ for $m > M$; Figure 7.2 shows the curve for $M > m$. The combined reliability curves are obtained by multiplication of the curves of probability of surviving chance ($e^{-\lambda T}$) and probability of surviving wearout (R_w).

The figures show that up to the age T_1 the combined reliability curve follows the exponential curve. Later, wearout failures take over and the combined curve drops fast towards zero. Figure 7.1 is much more characteristic of component behavior than Figure 7.2.

Equation (7.1a) can be used only for the cumulative probability from $T = 0$, when the component is new. If the component has already accumulated some age T_0—for instance, half of its useful life—Equation (6.23) for conditional reliabilities must be used:

$$R(t) = e^{-\lambda t}\, \frac{R_w(T_0 + t)}{R_w(T_0)} \tag{7.2}$$

Its value equals 1 for $t = 0$, i.e., for the beginning of the operating interval under consideration. In Figure 7.3 the curve from Figure 7.1 is replotted by the use of Equation (7.2) for an operation after the component has reached age T_0.

Originally, when the component was new ($T = 0$), the combined curve of the example in Figure 7.1 followed the exponential curve for about four-ninths of M hours. After the component has reached an age of about $T_0 = 0.4M$, the combined curve in Figure 7.3 follows the exponential curve for only a small fraction of the original M hours after T_0 and will drop rapidly towards zero after the time $t = T_2$, i.e., when a component operational age of $T_0 + T_2$ has been reached. The dotted curve shows the original position of the combined reliability curve with reference to the exponential curve.

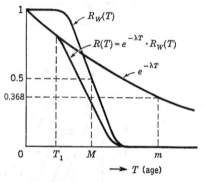

Fig. 7.1. Reliability curve for $m > M$.

It can be noted that whereas the position of m (mean time between failures) has not changed, the position of M (mean wearout life) has shifted much closer to the origin of the new coordinate system because of the component's age. It also becomes obvious that when the component exceeds the age $T_0 + T_2$, it will no longer behave exponentially and will become extremely unreliable.

This change, the transition from exponential behavior to a wearout behavior, occurs for a single component generally at an age between $M - 3.5\sigma$ and $M - 3\sigma$. For lower ages ($M - 4\sigma$, and less) the probability of wearout failure becomes insignificant for a single component.

But when many components are operating in series, even these extremely small probabilities of wearout failure of the individual components combine to significant figures. This situation was discussed in Chapter 6, and Equation (6.24),

$$R_s(t) = \exp\left[-\Sigma\,\lambda_i t\right] \cdot \Pi\, \frac{R_{wi}(T_i + t)}{R_{wi}(T_i)}$$

gives the resulting system reliability. The T_i values represent the ages of the various components in the system.

Figure 7.4 shows graphically the effect of wearout on the reliability of a multicomponent system as compared to the reliability curve of a

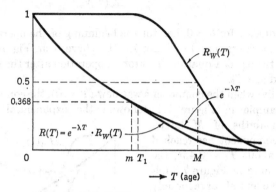

Fig. 7.2. Reliability curve for $M > m$.

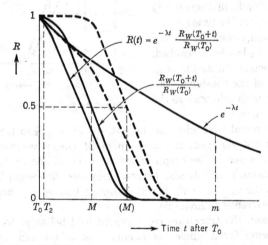

Fig. 7.3. Reliability of a component after it has reached an age T.

single component. Assumed is an electronic system of $i = 1000$ equal or very similar components in series, each of which has $m = 1,000,000$ hours, $M = 10,000$ hours, and $\sigma = 2000$ hours. The system begins to operate at the time $T_i = 0$, when all components are new and in perfect condition (i.e., assume that early failures have been removed by debugging). Therefore Equation (6.24) assumes the form

$$R_s = e^{-i\lambda t}[R_w(t)]^i \tag{7.3}$$

at this point. The curves do not show any spectacular effect of wearout on system reliability in the first several hundred hours of system operation. It would appear that, up to a system age of about 600 or 700 hours, only chance failures will occur, and there will be no wearout failures at all. Figure 7.4 shows how system reliability is reduced as compared to the reliability of a single component, but this is usual with components in series. Because the effect of wearout on the system reliability R_s appears to be comparatively small, one might easily decide to operate this system so that components are replaced only as they fail. However, this would be disastrous for the system's reliability beyond about the first thousand hours of the system's operational life.

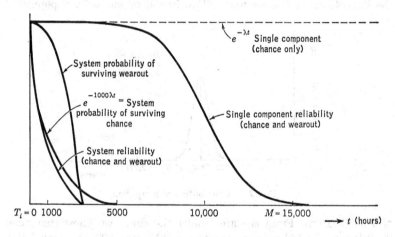

Fig. 7.4. Component and system reliability.

The failure rate of the system would rapidly increase and the system would prove extremely unreliable even for short operating periods. There would be extreme fluctuations of very high failure rates, until after a few thousand hours the failure rate would settle to an approximately constant figure of one failure every hour. Whereas during the first 600 or 700 hours of the system's operation the mean time between failures of the system was $m_s = m/1000 = 1000$ hours, system mean time between failures has now dropped to 1 hour. This means that the system's failure rate has increased 1000 times. Whereas at the beginning the system was capable of operating with a reliability of 0.99 for 10 hours a day, its operational capability is now down to 0.37 for a single hour and the system becomes incapable of a continuous operation.

Imagine, for instance, that this were the case with our radar defenses. Work out for yourself what would happen if there were not 1000 but, say,

10,000 components in series in a system and that they were replaced only when they failed. What operational reliability could be expected from such systems should an emergency arise?

We shall now show graphically what happens when components are replaced only "as they fail."* For this purpose we consider an example of 10,000 incandescent lamps or any other components with an assumed mean wearout life of $M = 7200$ hours and a standard deviation of $\sigma_1 = 600$ hours. Let us exclude chance failures and early failures. Thus only wearout failures are under consideration, and we shall see how the failure rate will fluctuate at first before it settles to a constant value.

Figure 7.5 illustrates the normal or Gaussian distribution curve of the lamp lives, or the wearout failure frequency curve for a population

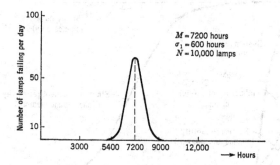

Fig. 7.5. Distribution of lamp life.

of 10,000 lamps. From the area under the curve we know that about 9970 lamps will fail between the hours 5400 and 9000, the peak failure period being around 7200 hours. By replacing each lamp as it fails we keep 10,000 lamps in operation continuously, for 10 hours per day.

When the first generation of lamps starts to wear out, the second generation of lamps begins to enter service from about 5000 hours. The lamps of the second generation are not introduced into service simultaneously, but gradually, as those of the first generation fail. Therefore, the wearout curve of the second generation of lamps will be considerably flattened, as illustrated in Figure 7.6.

The wearout peak of the second generation of lamps occurs at the time $2M = 14,400$ hours but this peak in our large sample is about halved against the failure peak of the first generation, and the standard deviation has approximately doubled to $\sigma_2 = 2\sigma_1 = 1200$ hours. From about 10,000 hours on, the third generation of lamps begins to be introduced.

* Igor Bazovsky, "Chance and Wearout Failure Rates," *Electronic Equipment Engineering*, 8:3, 113–15 (March, 1960).

It reaches its wearout peak at $3M = 21{,}600$ hours and its failures are distributed in time with a standard deviation of approximately $\sigma_3 = 3\sigma_1 = 1800$ hours. Because of the wider distribution, the failure peak is again lower—about one-third of the first peak. It can be seen from Figure 7.6 that to obtain the number of lamps N_F failing per unit time, we have to add the failure frequency curves of the second and third generation. This happens in our example from about 14,400 hours onwards. Later, from about 19,000 hours, we must also add the failures of the fourth generation which started to enter service as the third generation of lamps began to wear out. From that time on the number of lamps failing per unit time remains essentially constant, and the mixed age lamp population has

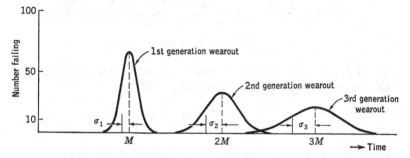

Fig. 7.6. Wearout curves of three lamp generations.

assumed a constant failure rate of $\lambda_r = 1/M = 0.000139$ per unit hour. The whole system of 10,000 lamps then has a failure rate of

$$\lambda_s = N \cdot \lambda_r = 10{,}000 \cdot 0.000139 = 1.39$$

lamps failing per hour. Because this constant failure rate of the mixed lamps results from wearout exclusively and because it represents the rate at which the lamps are replaced, we call it the *wearout replacement rate* (λ_r). It settles to an almost constant value shortly after the stabilization time $T = nM$, where $n = M/3\sigma_1$. In our example $n = {}^{7200}\!/_{1800} = 4$, and therefore the lamp wearout replacement rate can be considered as constant for all practical purposes from the time $T = 4M = 28{,}800$ hours.

The addition process and the stabilization of the failure frequency are shown in Figure 7.7 in a compressed scale by the plain line.

We have shown in this example that components stemming from a typical normally distributed population with a mean wearout life M and a standard deviation σ assume, in a system in which components are replaced as they fail, an essentially constant failure rate of $\lambda_r = 1/M$ per unit hour when mixed by age. This mixing process is essentially completed after the stabilizing time $T = M^2/3\sigma$. The mean wearout life M then equals the mean time between failures m of the mixed age components. Because of the constant failure rate, failures will now occur at random time intervals, but they are not chance failures—they are purely

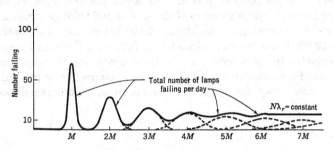

Fig. 7.7. Stabilization of failure frequency.

wearout failures. With 10,000 of these components "in series" in a system, the system mean time between failures will be

$$m_s = \frac{m}{10,000}$$

and the system's probability of survival, or reliability, will be

$$R_s = e^{-t/m_s} = e^{-10,000t/m}$$

With a component wearout failure rate of 0.000139, such as that of the lamps, the system mean time between failures would be only

$$m_s = \frac{7200}{10,000} = 0.72 \text{ hours}$$

This would result in an extremely low reliability. But the system has assumed a constant failure rate and will behave exponentially despite that only wearout failures of components occur.

The same deliberations made above apply to any mixed age population of components, whatever their underlying distribution may be.* And if components are mixed by types from the beginning, each type having a different mean wearout life M, or a different distribution, the system failure rate stabilization process progresses relatively much faster than in this lamp example. The mixing by types is in fact always present in all equipments and systems.

* R. von Mises, *Wahrscheinlichkeit, Statistik und Wahrheit* (*Probability, Statistics and Truth*), 3rd ed., Springer-Verlag, Vienna, 1951.

The question is whether reliable systems can be built from 1000, 10,000, or even more components such as the lamps, when all are put in series. It will now be shown that, under certain circumstances, even extremely high system reliabilities can be achieved.

Until now, in this example, components were replaced only "as they failed." They were allowed to operate until worn out, as is often done with electronic and other components. The consequence is an intolerably high failure rate.

To reduce the constant wearout failure rate of the system, the components must be replaced *before* they wear out. This increases system reliability by several orders of magnitude. Had we adopted a policy of replacing all 10,000 lamps in our example at a time of $M - 6\sigma_1 = 3600$ hours, none of the lamps would have failed in the period from 3600 to 7200 hours. And if we had replaced all lamps again at 7200 hours, and again at 10,800 hours, etc., we could have had an almost failure-free system. In fact, we would expect not more than a single lamp to fail out of 10,000 lamps in 100,000 hours of system operation. The system reliability for any period of 3600 system operating hours between the replacements would amount to 0.99999, and correspondingly much higher reliabilities would be obtained for short stretches of, say, 10-hour missions. The exact figures can be derived from the appropriate areas under the normal distribution curve of the first generation of lamps and from consideration of the number of components in the system.

The increases in system reliability are striking when compared with the extremely low reliability obtained when the lamps were replaced only as they failed. System mean time between failures was 0.72 hours but by adopting a strict preventive replacement schedule every 3600 hours, we improved the system mean time between failures to 100,000 hours—an increase by five orders of magnitude.

To sum up, when chance failures are absent so that only wearout failures occur and components are replaced only as they fail, such series system assumes, after a stabilization period during which high failure rate fluctuations occur, a constant failure rate of

$$\lambda_s = \sum_{i=1}^{n} \frac{1}{M_i} \tag{7.4}$$

where M_i is the mean wearout life of the ith component. System reliability is then

$$R_s = e^{-\lambda_s t} = \exp\left[-\sum_{i=1}^{n} \frac{1}{M_i} t \right] \tag{7.5}$$

because, after stabilization, the system behaves exponentially although the individual components fail because of wearout. The failure density curve of each component is normal and not exponential. Measured thus in a time scale of the component's life, wearout failures cluster around its mean wearout time and are distributed normally. But measured in the system's time scale, these component wearout failures occur at random time intervals and are distributed exponentially—because once the stabilization process is completed, the individual components enter their service in the system at random times in the system's time scale and, consequently, they will also fail at random times from the system's point of view.

The addition in Figure 7.7 is done by adding up the instantaneous values of the frequency functions of those generations of lamps which are operating at a given system time. But because the frequency function of a generation of lamps is but the sum of the density functions of the individual lamps of that generation, the addition process which yields the instantaneous value of the system's failure rate—the plain line in Figure 7.7—can also be carried out by summing the instantaneous values of the density functions of all individual components operating at a given system time. This procedure is used when the components are not equal but have different means and different distributions with different variances.* However, this is of interest only if we want to know exactly the transient situation which prevails before the system failure rate stabilizes to its final value as given by Equation (7.4). Therefore Equations (7.4) and (7.5) are always valid for a stabilized system failure rate when components are replaced only as they wear out and regardless of the density function of the individual components of which the system is composed.

A different situation prevails when the components in a series system are replaced preventively before they get a chance to fail because of wearout. The reliability of this system is given for a system operating period t by

$$R_s(t) = \prod_{i=1}^{n} \frac{R_{wi}(T_i + t)}{R_{wi}(T_i)} \tag{7.6}$$

when no chance failures occur and only the probability of wearout failures is considered. The right side of this equation is the product of the reliabilities of all n components in the system. The reliability of the ith component for a system operating time t is derived from Equation (6.22):

$$R_{wi}(t) = \frac{R_{wi}(T_i + t)}{R_{wi}(T_i)} = \frac{\int_{T_i+t}^{\infty} f_i(t)\, dt}{\int_{T_i}^{\infty} f_i(t)\, dt} \tag{7.7}$$

* Dr. Erich Pieruschka "Mathematical Foundation of Reliability Theory," Redstone Arsenal, Huntsville, Alabama (January 1958).

where $f_i(t)$ is the density function of the component and T_i is the age to which this component has lived since it was put into service, i.e., since the last replacement up to the beginning of the operating period t under consideration.

Because no component remains in service longer than up to its replacement time T_{wi}, the minimum system reliability becomes

$$R_s(t) = \prod_{i=1}^{n} \frac{R_{wi}(T_{wi})}{R_{wi}(T_{wi} - t)} \tag{7.8}$$

When all components in the system are replaced simultaneously at a replacement time T_w, Equation (7.8) results in

$$R_s(t) = \prod_{i=1}^{n} \frac{R_{wi}(T_w)}{R_{wi}(T_w - t)} \tag{7.9}$$

If chance failures are also present along with wearout, the stabilized system failure rate, when components are replaced only "as they fail," is

$$\lambda_s = \sum_{i=1}^{n} \lambda_{ci} + \sum_{i=1}^{n} \frac{1}{M_i} \tag{7.10}$$

and the reliability of this system, after the failure rate has stabilized to the above value, is

$$R_s(t) = \exp\left[- \sum_{i=1}^{n} \left(\lambda_{ci} + \frac{1}{M_i}\right) t \right] \tag{7.11}$$

where M_i is the mean time between wearout failures of the ith component when chance failures occur at a rate of λ_{ci}. If we use the designation $\lambda_r = 1/M$ for the wearout replacement rate and λ_c for the chance failure rate, the component assumes in the system a final failure rate of $\lambda = \lambda_c + \lambda_r$, and Equation (7.11) can be written as

$$R_s(t) = e^{-\sum_{}^{n}(\lambda_c+\lambda_r)_i t} = e^{-\sum_{}^{n}\lambda_i t} \tag{7.12}$$

The system failure rate $\sum_{i=1}^{n} \lambda_i$ is then a constant and is the resulting sum of the constant failures rates of the n individual components which comprise both chance and wearout failures.

Equation (7.12) is deceptively similar to the purely exponential case where only chance failures occur. It therefore easily escapes attention that the λ_r term can be eliminated from the constant failure rate by timely component replacement. As we shall see in the next chapter, there is often also a third constant failure rate added to λ_c and λ_r, that resulting from early failures.

A warning must be given here that component failure rates derived from observing systems in operation for extended periods of time, although they appear to be constant, are not necessarily caused by chance failures at all, and may very well be the result of wearout only. Thus, constant failure rates are not a criterion for distinguishing between wearout and chance failures. Only careful statistical analyses and tests can discover the pure chance failure rates of components. Component failure data taken from system observations are normally a mixture of early, chance, and wearout failures which are not easy to separate.

If high system reliabilities are required for extended periods, component wearout failures must be categorically prevented by early replacement of each component which is known to be subject to wearout, and only components free from early failures may be used as replacements. When this rule is followed and when properly rated reliable components are used, extremely high system reliabilities can be achieved.

Chapter 8

EARLY FAILURES AND THE LIFE
FUNCTION OF COMPONENTS

THE IMPORTANCE OF A STATISTICAL TREATMENT, combined with the calculation of the probabilities of chance and wearout failures, for the evaluation of component and system reliabilities has become obvious in the preceding chapters. In fact, the statistical probability approach is the only possible approach to reliability engineering when the equipment must operate under high stresses in a difficult environment and when weight considerations do not allow the application of very high safety factors.

In general, reliability problems can be well treated mathematically by means of the exponential and normal failure distributions.

In two extreme cases, the exponential distribution alone is sufficient for calculating reliabilities:

1. when wearout is systematically prevented by replacing components before they reach the end of their useful life,
2. when components are allowed to wear out and are replaced only as they fail.

We shall see that almost all reliability evaluations which are concerned with calculating the probability of completing successfully a given mission fall into these two categories. In case 1, reliabilities higher by several orders of magnitude can normally be achieved than in case 2. Whereas case 2, from the time the failure rate has stabilized, is strictly exponential, in case 1 wearout studies are involved to determine the best preventive replacement time for each component. Also, whereas case 2 applies only to long-life equipment which is subject to repair maintenance during its operational life, case 1 applies to both long-life and one-shot equipment thus:

(a) to long-life equipment when high reliabilities are required in operation and preventive maintenance is routine,
(b) to one-shot equipment whose operating period is so short that no component can possibly wear out and therefore the combined wearout probabilities of large numbers of series components can be neglected.

To assure that this can be done, the wearout characteristics of the components must be known. So, a true reliability program cannot neglect wearout studies and concentrate only on chance failure studies. Wearout studies are absolutely essential for the establishment of preventive maintenance policies (long-life equipment), or for assuring that series wearout phenomena cannot affect a system during a mission (one-shot equipment).

Besides chance and wearout failures, there are the early failures which can also adversely affect the reliability of a system. The probabilistic treatment of this type of failure is less important than that of the chance and wearout failures. Usually, the early failures are eliminated within the first 20 to 200 hours of component and system life by means of systematic burn-in or debugging procedures, and they should not affect reliability when debugging is completed and the system goes into operational service. The time it takes to eliminate early failures is a matter of engineering experience, but for the benefit of gaining a complete picture of the whole reliability problem, the early failure phenomenon will now be discussed.

We said earlier that chance failures were defined as failures caused by unpredictable sudden stress accumulations outside and inside of the components beyond the component design strength. Sudden stress accumulations are unavoidable in airborne and similar equipment; stress accumulations occur at random time intervals, at various stress levels, in various combinations of the multitude of different stresses, and at various locations in an equipment.

A component fails when stress combinations or individual stresses exceed its strength. If in a component lot there are a few initially weak, substandard specimens, these will naturally fail at much lower stress levels than the rest. But the mechanism of component early failures is still the same chance mechanism which causes all chance failures—only the low levels of stress accumulations occur much more frequently than the extreme stress accumulations. Therefore, the substandard components have, as a separate lot, a much higher failure rate—or a much lower mean time between failures—than the good components.

The substandard components usually have their own very high failure rate, which, however, is constant because of the chance mechanism involved. Their failures are thus distributed exponentially with a mean time between failures smaller by several orders of magnitude than the mean time between failures of the rest of the population. The weak specimens in a large population of good components, being replaced by good components when they fail, die out exponentially and very fast. Because no advance knowledge exists as to which are the substandard specimens, their presence affects the reliability and the failure density function of the whole lot of components. This is consistent with the definition of reliability of a lot of components, $R(t) = N_s/N$, where N_s is the number of survivors at time t from an initial total lot of N components. Obviously, if there are N components in series in a system, and one or a few of these are substandard specimens, the system's reliability is almost exclusively determined by the extremely high probability of failure of these few specimens. The system will be unreliable as long as these weak specimens are not eliminated.

When substandard components are present in the initial stages of operation, system reliability is very low; but it improves rapidly as the failed weak components are replaced by good ones.

To investigate the effect of substandard components on system reliability, let us assume a system of N components of which N_E are substandard and let $N_E \ll N$. Also let us at first assume that the good components are 100 per cent reliable, whereas the weak components have a very short mean time between failures of m_e each. The probability of failure of a weak component is then $Q = 1 - e^{-t/m_e}$, and the probability of failure of N_E weak components in series is $Q = 1 - e^{-N_E t/m_e}$. Because the good components have been assumed 100 per cent reliable, the reliability of the whole series system of N components in series of which N_E are substandard amounts to

$$R_s = e^{-N_E t/m_e}. \qquad (8.1)$$

Thus, the reliability of a system when it is first put into operation is completely determined by the unreliability of the N_E substandard components. Because the mean time between failures of these components is usually very low, from several minutes to several dozen hours, the system is very unreliable. Now, as failures occur, the population of the substandard components N_E decays exponentially, while N remains constant because the failed weak components are replaced by good ones. By the same token, system reliability improves or grows with each replacement. After the first failure and its repair, system reliability becomes $R =$

$e^{-(N_E-1)t/m_e}$, after the second failure $R = e^{-(N_E-2)t/m_e}$, etc. But as long as even one substandard component remains in the series system, system reliability cannot be better than the reliability of this single component which is $R = e^{-t/m_e}$. And because m_e is so short, the system remains extremely unreliable. However, as soon as this last substandard component fails and is replaced by a good one, system reliability suddenly jumps to 100 per cent, because we assumed a zero failure rate for the good components.

For instance, if there are 10 weak components in such a system and each has $m_e = 10$ hours, system reliability before the first failure is $R = e^{-1} = 0.368$ for a 1-hour operation and $R = e^{-10} = 0.0000454$ for a 10-hour operation. After the ninth failure, when only one weak component is left in the system, system reliability becomes $R = 0.9$ for 1 hour and $R = 0.368$ for 10 hours. But as soon as the last weak component fails and is replaced, system reliability becomes $R = 1$ for 10 hours and also for any other operating period, when we assume that all good components are 100 per cent reliable. The debugging of a system thus consists of operating the system under environmental conditions similar to or more severe than those which will prevail in actual service to bring the weak components to failure so they can be replaced by good ones.

In debugging procedures we are usually interested in the length of the debugging time, which means: how long does it take a system to achieve its useful life reliability? If there are initially N_E substandard components in the system, what period of time elapses before these fail? This is a statistical question which can be answered in terms of the expected time for N_E components to fail, if the mean time between failures of each of these components is m_e. Obviously, the expected time to failure of one component is m_e, and therefore, when there are initially N_E of these components, the expected time for all of them to fail is:

$$E(t) = m_e\left(1 + \frac{1}{2} + \frac{1}{3} + \cdots + \frac{1}{N_E}\right) \qquad (8.2)$$

This formula is derived in Chapter 11, Equation (11.23). A formula for $E(t)$ when the mean times to failure m_e of the components are not identical is obtained by extending Equation (11.21) in the same chapter. We can refer to $E(t)$ as the *mean debugging time*. For the 10 substandard components with $m_e = 10$ hours each in the abovementioned example, the expected time for the system to be debugged is, according to Equation (8.2), about $E(t) = 3m_e$, or 30 hours. If there were 100 of these substandard components, the expected time would be about $E(t) = 5.5m$ or 55 hours.

However, it is important to realize that $E(t)$ is only an average time in the sense of an exponential mean time between failures, which means that only about 63 out of 100 such systems would be debugged in the expected time $E(t)$. Or, stated differently, in an average system one would expect 63 per cent of the potential early failures to have occurred in the period from time $t = 0$ to the time $t = E(t)$. Therefore the mathematical reliability of any one system at an age $E(t)$ is still very low, about $R = 0.67$ for a 1-hour operation, and about $R = 0.045$ for a 10-hour operation, as will be shown later. Thus, to achieve high system reliability in operation, the actual debugging time should extend to at least five times the mean debugging time, i.e., $5E(t)$.

The knowledge that early failures are distributed exponentially and that they decay exponentially will allow us to draw conclusions as to the required actual debugging time of the system from observing the time to

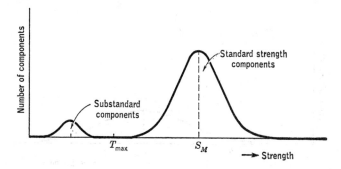

Fig. 8.1. Strength distribution of standard and substandard components.

failure of the first few early failures which occur in the debugging process.

What we have said about early failures so far was based on three assumptions: (1) that a separate substandard population exists in an otherwise good population of components of standard strength; (2) that only good components are used to replace failed substandard components; and (3) that the failure rate of the good components is zero. These three assumptions will now be discussed to see what happens if they do not hold.

Assumption (1), the existence of a separate substandard population in an otherwise good population of standard strength, can be depicted graphically by the distribution of strengths of the two populations, as shown in Figure 8.1. $T_{\max}$ in this graph is to illustrate the maximum occurring stress accumulation in operation. This case of component distribution occurs in practice when the variance of component strength is relatively small, but because of some imperfection in the control of the production process, a small proportion of components is not up to the over-all standard.

Another case of early failures arises when the variance of component strength is large and the quality of the produced components is not at all uniform to begin with. The strength distribution of such a component population could look as shown in Figure 8.2. From a reliability point of view, the use of such components is most undesirable. Such populations resist debugging procedures and maintain consistently high failure rates.

Comparing the two cases, let us assume that the population of good components in the first case and the population of the components with a wide variance in the second case have the same mean strength S_M. Also, let the same stress spectrum with a T_{max} maximum stress accumulation act on both populations. Clearly, whereas in the first case the population of good components is, for all practical purposes, immune to the occurring stresses, and therefore the system is almost absolutely reliable once the

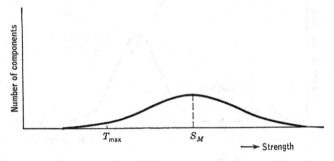

Fig. 8.2. Components with nonuniform strength distribution.

substandard components have been replaced by good ones, in the second case a high probability of failure will persist at the T_{max} stress point. The use of component populations as shown in Figure 8.2 is to be avoided when high system reliabilities are required.

Assumption (2), where every failed substandard component is replaced by a good one, is called "perfect repair" because no new element of early failure is introduced into the system when failed components are replaced or when other repair action is taken. Because early failures can be caused not only by substandard components but also by poor assembly practices, N_E can be the number of any initial weak spots in the system or, in general, the number of potential early failures.

If repair is not perfect, a certain proportion of new early failures is introduced as failures are repaired. These newly infiltrated, early failures may stem from the introduction of some substandard components into the system when failed components are being replaced or from other imperfections of the repair action (for instance, poor solder joints and other mistakes). Thus, whereas in the case of perfect repair no potential early failures remain in the system after the original N_E potential early failures have been realized and repaired, in the case of nonperfect repair there will still be potential early failures in a system, when the original ones have been eliminated.

Assume that the cause of imperfect repair is a substandard component only and not any other repair mistake, and that the replacement components are taken from a population in which the proportion of substandard specimens is N_B/N in a lot of N components. Now, as the original N_E weak components in the system have failed and been replaced, on the average

$$N_2 = N_E - N_E \cdot \left(1 - \frac{N_B}{N}\right) = N_E \cdot \frac{N_B}{N} \qquad (8.3)$$

new potential early failures have been introduced into the system by the nonperfect repair procedure. And if the replacement components are taken from a population which contains the same proportion of substandard specimens as the population from which the system was built, i.e., N_E/N, then, by the time the original N_E weak components in the system have failed, an average of

$$N_2 = N_E - N_E \left(1 - \frac{N_E}{N}\right) = \frac{N_E^2}{N} \qquad (8.4)$$

new potential early failures have been introduced.

When the N_E^2/N components fail, the expected number of early failures introduced because of these components will be only

$$N_3 = N_2 \frac{N_E}{N} = \frac{N_E^3}{N^2} \qquad (8.5)$$

Thus, the system debugging proceeds slower now than in the case of perfect repair; but when only one potential early failure remains in the system, the probability that a good component will be picked for its repair when this last early failure occurs is $1 - N_E/N$. This is normally a high enough probability and, therefore, with nonperfect repair the system will become 100 per cent reliable when we assume that the good components have zero failure rates.

This brings us to the last assumption (3). To assume that the good components have a zero failure rate is an oversimplification of the problem. Let us see what happens when the good components also have some constant chance failure rate, although it may be very low.

The effect of substandard components or other potential early failures in such a population of predominantly good components becomes quite different from the previous case where the good components were assumed to have zero failure rates. When a system consists initially of N_G good components with a failure rate of λ_g each and N_E substandard components with a failure rate of λ_e each, and $\lambda_e \gg \lambda_g$, the system has an initial system failure rate of

$$\lambda_{si} = N_G\lambda_g + N_E\lambda_e \tag{8.6}$$

where $N = N_G + N_E$ is the total number of components in the system.

With a perfect repair policy, the N_E components die out exponentially as shown before, and the system failure rate stabilizes to

$$\lambda_s = N\lambda_g \tag{8.7}$$

as soon as the last substandard component has failed and has been repaired. Therefore, system reliability which was low initially because of the presence of substandard components or other potential early failures grows as the debugging proceeds to reach a much higher operational reliability defined by the constant system failure rate in Equation (8.7).

However, when repair is not perfect, the system will not achieve the reliability defined by (8.7) because a fraction of potential early failures will trickle into the system all the time. The initial failure rate will still be that given by Equation (8.6), will also drop rapidly as the debugging of the system proceeds, but will finally stabilize to a value of

$$\lambda_s = N\lambda_g \frac{n}{n_G} \tag{8.8}$$

where n_G is the number of good repair actions out of a total of n repair actions, or, in other words, when n repair actions result in the introduction or infiltration of $n - n_G = n_E$ early failures into the system.

Equation (8.8) can be rewritten making use of the identity $n = n_G + n_E$ as follows:

$$\lambda_s = N\lambda_g \left(1 + \frac{n_E}{n_G}\right) \tag{8.9}$$

where n_E/n_G is the ratio of substandard components to good components in the component population from which the replacements are taken. Or, in general, it is the ratio of poor repair actions which result in early failures to good repair actions.

This ratio n_E/n_G is an important factor in cases where nonperfect repair is involved, because it determines the increment of the stabilized system failure rate over the stabilized system failure rate with perfect repair which was $\lambda_s = N\lambda_g$. We call n_E/n_G the incremental failure rate factor.

Equation (8.8) can also be written as

$$\lambda_s = \frac{N\lambda_g}{n_G/n} \tag{8.10}$$

by putting the reciprocal value of n/n_G into the denominator. This reciprocal value n_G/n is then called *repair efficiency*.*

The graph in Figure 8.3 shows the process of system failure rate stabilization with perfect and with nonperfect repair. In the case of nonperfect repair the failure rate stabilization process may assume a damped oscillatory character when $N_E \ll N_G$ in the initial system condition and when $\lambda_e \gg \lambda_g$.

Finally, we want to look into the reliability of systems which are not subject to repair once they are put into operation or when they take off, such as missiles and space vehicles. For this purpose we need to know the failure rate and reliability of the components, both of which are now functions of component age.

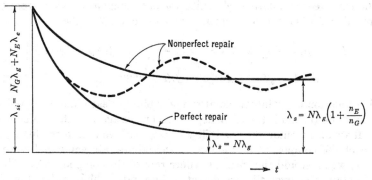

Fig. 8.3. Failure rate stabilization with perfect and imperfect repair.

When, in a population of components, there are N_E substandard specimens with a failure rate of λ_e, and N_G good components with a failure of λ_g, so that, regardless of the size of the lot from which the components are taken, the percentage of the substandard specimens is

$$\frac{N_E}{N_E + N_G} = \frac{N_E}{N} \cdot 100\%$$

the expected failure rate of any of the components picked from that lot, without the knowledge of whether or not it is substandard, becomes a function of the age T of the component and amounts to

$$\lambda = \lambda_g + \lambda_0 e^{-T/E(t)} \tag{8.11}$$

where $\lambda_0 = (\lambda_e - \lambda_g)N_E/N$ and $E(t)$ is defined by Equation (8.2). Then,

* Dr. Erich Pieruschka, "The Theoretical Foundation of the Maintenance of Complex Equipment," Redstone Arsenal, Huntsville, Alabama (June 1958), p. 11.

at time $T = 0$ when the component is new, its expected initial failure rate is

$$\lambda_i = \lambda_g + \lambda_0 = \lambda_g + (\lambda_e - \lambda_g)\frac{N_E}{N}$$

$$= \lambda_g\frac{N - N_E}{N} + \lambda_e\frac{N_E}{N} = \frac{\lambda_g N_G + \lambda_e N_E}{N} \qquad (8.12)$$

As the component operates and if it does not fail in the initial stages of operation, its failure rate—which means its instantaneous probability of failure—rapidly converges to a final value which we obtain from Equation (8.11) by taking the limit as T increases indefinitely:

$$\lambda = \lambda_g \qquad (8.13)$$

The expected cumulative reliability for one component taken from the time $T = 0$ when the component is new to an age T is then

$$R = \exp\left[-\int_0^T \lambda\, dT\right] = \exp\left[-\int_0^T (\lambda_g + \lambda_0 e^{-T/E(t)})\, dT\right]$$

$$= \exp\left[-\lambda_g T - \lambda_0 E(t) + \lambda_0 E(t)e^{-T/E(t)}\right] \qquad (8.14)$$

This equation takes into account the possibility of early failure as well as chance failure, and is therefore dependent on the age T of the component.

If we also include the possibility of wearout, we obtain the life equation of the component as follows: to the failure rate value in Equation (8.11) we must add the wearout failure rate of the component λ_w. Thus, we obtain the over-all component failure rate which includes early, chance, and wearout failure possibilities:

$$\lambda_L = \lambda_g + \lambda_0 e^{-T/E(t)} + \lambda_w \qquad (8.15)$$

According to Equation (4.14) and Figure 6.5, the wearout failure rate of a component is defined as

$$\lambda_w = \frac{f(T)}{R(T)} = \frac{\exp\left[-\dfrac{(T - M)^2}{2\sigma^2}\right]}{\displaystyle\int_T^\infty \exp\left[-\dfrac{(T - M)^2}{2\sigma^2}\right] dT} \qquad (8.16)$$

in the case of the Gaussian distribution of wearout failures. If it is found in some specific case that a different type of distribution gives a better fit for the wearout phenomenon, for instance, the log normal distribution, then the wearout failure rate λ_w is computed from the density function $f(T)$ of that particular distribution as

$$\lambda_w = \frac{f(T)}{R(T)} = \frac{f(T)}{\displaystyle\int_T^\infty f(T)\, dT} \qquad (8.17)$$

The total failure rate λ_L is thus a function of the age T of the component:

$$\lambda_L = \lambda_g + \lambda_0 \cdot e^{-T/E(t)} + \frac{f(T)}{\int_T^\infty f(T)\,dT} \tag{8.18}$$

The cumulative reliability, or the life function $L(T)$, of the component, counted from the time $T = 0$ when it first goes into operation, is then

$$L(T) = \exp\left[-\int_0^T \lambda_L\,dT \right] \tag{8.19}$$

$L(T)$ is also the cumulative probability that the component will be alive at an age T.

The reliability of the component for an operating interval t (for instance, for a mission of length t), when at the beginning of that interval the age of the component is T, becomes

$$R(t, T) = \exp\left[-\int_T^{T+t} \lambda_L\,dT \right] \tag{8.20}$$

Only when there is absolute assurance that the component cannot become an early failure, i.e., that it was taken from a perfect lot, and that in the period t under consideration it cannot suffer any wearout deterioration, will Equation (8.20) reduce to an exponential form, independent of the age:

$$R(t) = \exp\left[-\lambda_g t\right] \tag{8.21}$$

However, since a large number of series components are involved in missile and space systems, even minute chances of early failure and wearout failure must be considered because of their additive effect over all system components.

Equations (8.19) and (8.20) can also be written in the form of a product of the component's probability of surviving chance, infantile weakness, and wearout:

$$L(T) = R_c(T) R_e(T) R_w(T)$$

$$= \exp\left[-\lambda_c T\right] \cdot \exp\left[-\frac{\lambda_0}{\alpha}(1 - e^{-\alpha T}) \right]$$

$$\times \exp\left[-\int_0^T \frac{f(T)}{\int_T^\infty f(T)\,dT}\,dT \right]^* \tag{8.22}$$

and

$$R(t, T) = R_c(t) R_e(t) R_w(t)$$

$$= \exp\left[\lambda_c t\right] \cdot \exp\left[-\frac{\lambda_0}{\alpha} e^{-\alpha T}(1 - e^{-\alpha t}) \right]$$

$$\times \exp\left[-\int_T^{T+t} \frac{f(T)}{\int_T^\infty f(T)\,dT}\,dT \right]^* \tag{8.23}$$

* $f(T)$ is the wearout failure density function.

The above equations are derived from Equation (8.14) by its multiplication by the probability that no wearout failure will occur. We substituted λ_c, the chance failure rate, for λ_g, and α for $1/E(t)$ so that α is the reciprocal of the mean debugging time. The wearout density function $f(T)$ where T is the operating age of the component can be Gaussian normal, log normal, Weibull, or of any other form.

In the case of a Gaussian normal distribution of wearout failures, we make use of Equation (7.1b) and obtain for the life function:

$$L(T) = \exp\left[-\lambda_c T\right] \cdot \exp\left[-\frac{\lambda_0}{\alpha}(1 - e^{-\alpha T})\right] \cdot \frac{1}{\sigma\sqrt{2\pi}} \int_T^\infty e^{-(T-M)^2/2\sigma^2}\, dT$$

$$(8.24)$$

and, making use of Equation (6.21), the reliability at an age T becomes

$$R(t, T) = \exp\left[-\lambda_c t\right] \cdot \exp\left[-\frac{\lambda_0}{\alpha} e^{-\alpha T}(1 - e^{-\alpha t})\right] \cdot \frac{\displaystyle\int_{T+t}^\infty e^{-(T-M)^2/2\sigma^2}\, dT}{\displaystyle\int_T^\infty e^{-(T-M)^2/2\sigma^2}\, dT}$$

$$(8.25)$$

When n components combine into a series system, the system's reliability is then

$$R_s(t) = \Pi\, R_i(t, T_i) \tag{8.26}$$

where $R_i(t, T)$ is the reliability of the ith component and is given by Equation (8.25) for a component age T_i.

Equation (8.25), along with the more general Equation (8.23), is the complete expression for the reliability of a component throughout its operational life, when it operates at a given stress level so that the parameters remain unchanged. It follows from this equation that component reliability is not only a function of the operating period t, but also a function of its age T. Because of this fact we can define more precisely the reliability of a component, taking into account its age and the conditional probability character of reliability:

> *The reliability of a component is its conditional probability of performing its function within specified performance limits at a given age for the period of time intended and under the operating stress conditions encountered.*

As previously mentioned, the reliability of a component can come very close to being a pure exponential function when good debugging procedures are used, and when the component is designed for a life much longer than its total operating life in the system where it will be used. In missile and space vehicle applications, this means that the component must come from a completely debugged (i.e., perfect) lot of components which have a mean wearout life M much longer than the operating life of the system, including system pretesting, check-out, and mission time, and which are of a uniform quality so that their standard deviation is small compared to their mean life.* When it is assured that these requirements are fulfilled, then Equation (8.21) is a good approximation for component reliability, and it can be used in conventional system reliability calculations.

* Exceptions to this rule are discussed in Chapter 17 in connection with in-service maintenance practices.

Chapter 9

THE EXPONENTIAL AND POISSON
DISTRIBUTIONS IN RELIABILITY

A THOROUGH UNDERSTANDING of the early failure and wearout failure phenomena discussed in Chapters 6, 7, and 8 is very important in reliability engineering. Early failures and wearout failures are dangerous enemies of system reliability.

If a system which contains potential early failures is prematurely put into service, i.e., before the early failures are given a chance to be weeded out, the system will almost certainly fail shortly after it begins to operate. This may happen in a few seconds or minutes, or in the first few dozen hours. When the system is a strictly series system, such failure may be catastrophic. Without proper debugging the reliability of a system will be low because it is governed almost entirely by the substandard components or other potential early failures, as long as even a single one remains in the system. In the preceding chapter it was also shown how nonperfect repair affects system reliability in its useful life period. Imperfect repair also introduces early failures into the system during the useful life period, and this will increase the system's failure rate. Perhaps without these constantly introduced early failures, the system could be close to 100 per cent reliable. Where reliability really matters, we cannot overemphasize the need for good debugging, for using only debugged or burned-in components for replacement, and for applying only perfect repair techniques.

The expense incurred by adopting these strict procedures turns out to be less than that caused by having products returned as unsatisfactory, by having to repair the equipment at distant locations, by losing an aircraft, or a missile right after launching, or even by losing a reputation. Proper debugging procedures and perfect repair techniques, which include the use of burned-in components, are a matter of pretesting, check-out, component preaging, and sound engineering judgment backed by the necessary knowledge of statistics and probability calculations.

Another great enemy of reliability is wearout. System pretesting and checkout is not effective here—it will not help to keep a system free from wearout failures. To cope with wearout requires accurate information on the wearout characteristics of the individual components and on the distribution of the wearout symptoms or wearout failures, and probability calculations are necessary to determine the combined wearout effects of the numerous components in a system. It was shown in Chapter 6 how even very small wearout probabilities of individual components can combine to an extremely high probability of system failure when many components are involved. But wearout can be prevented by timely component replacement. To determine the time when components must be replaced in a system is, as shown, a matter of the calculus of probabilities. The more complex a system is, the earlier components must be preventively replaced. If this is not done, system reliability becomes ridiculously low, and such systems never cease to be costly maintenance nightmares.

If debugging and wearout prevention is not observed, complex systems may begin to fail because of component wearout even before they are properly debugged. In such situations we find that because no systematic debugging effort was made, early failures are reintroduced at every occasion. Such systems, before their failure rates stabilize at a high constant value, will go through periods where the exponential decay of early failures combines with constant rate chance failures and with normally or log normally distributed wearout failures. This may result in extravagant failure density functions, so that even good statisticians can run into difficulties when trying to find the distribution which most closely fits such transient situations.

All the efforts expended in debugging and in preventing wearout are made for a single purpose, which is the goal of all reliability engineering: to keep the system throughout its operational life in a condition in which only chance failures can occur, and to reduce these to an absolute minimum by proper design techniques. Thus, a "must" for a reliable system is that during its operation it be, as far as possible, free from early failures as well as from wearout failures. Then the system will behave strictly exponentially; if failures still occur they will be caused by chance—that is, by sudden and unpredictable accumulations of various stresses beyond the design strength and application level of the components and parts which form the system. The exponential chance failure condition (in which reliability can be well controlled by reliability methods which will be expounded in the following chapters) can then persist almost indefinitely, when proper maintenance procedures are applied.

To be prepared for an understanding and application of the reliability methods explained in later chapters, methods which are the heart and the muscles of the theory and practice of reliability, the exponential distribution derived by the great French mathematician, Poisson, will first be briefly reviewed. This function holds a very special and prominent position in reliability calculations because it describes so well the behavior of components and systems in their useful life period, i.e., when they display approximately constant failure rates. Its great advantage over any other statistical distributions is the single parameter λ, or its reciprocal m, which fully and completely describes a given exponential distribution.

A further advantage of the exponential function is that it is independent of the age of a component or system as long as the constant failure rate condition persists. As we have seen in Chapter 7, systems which are composed of many components and which are repaired when they fail or on which regular maintenance is practiced, always assume a constant failure rate condition over sufficiently long periods of observation. This by itself is already sufficient justification for the predominant position held by the exponential distribution function in reliability engineering.

The exponential density function is given by

$$f(t) = \lambda \cdot e^{-\lambda t} \tag{9.1}$$

where λ is the constant failure rate. We can see that this function is uniquely and completely determined when the value of the single parameter λ is known. This fact makes reliability testing of exponential equipment comparatively simple because all that is needed is to determine the value of λ by a test. Other distributions usually have more than one parameter. For instance, the Gaussian normal function has two parameters M and σ, the log normal function has three parameters M, a, and σ, etc. In such cases two or three parameters must be determined by tests before these functions are completely determined.

We have also learned in the previous chapters that, in general, to obtain the cumulative probability of failure, or unreliability Q, we have to integrate the density function. In the exponential case this results in

$$Q(t) = \int_0^t \lambda e^{-\lambda t} \cdot dt = 1 - e^{-\lambda t} \tag{9.2}$$

which is the probability of failure in an operating period t. As was shown before, in the exponential case this probability is independent of the age of the equipment, because the values of the cumulative probability and of the conditional probability are identical for periods of equal length.

From the probability theory we know that because the probability of failure Q (unreliability) and the probability of survival R (reliability) are the probabilities of two complementary events (failure and survival) for the same operating period t, we can write

$$R(t) + Q(t) = 1$$

Therefore, the exponential reliability is given by

$$R(t) = 1 - Q(t) = e^{-\lambda t} \tag{9.3}$$

When the failure rate λ is known (and because this equation does not depend on the age of the equipment), the equipment's reliability for any operating time t can be immediately calculated from Equation (9.3). The operating time t may be of any length, and may be taken at any age of the equipment with the sole limitation that the failure rate must not change during that period. For a single component it means that its strength must not deteriorate and that it will operate in essentially the same environment.

If, during an operation, the environmental stresses change from one constant level to another constant level but the component strength remains the same, the component will exhibit a constant failure rate λ' for the operating time t' at the first stress level and another constant failure rate λ'' for the operating time t'' at the other stress level. For a total operating time $t = t' + t''$ the reliability of the component is then

$$R(t) = e^{-\lambda' t'} \cdot e^{-\lambda'' t''} = e^{-(\lambda' t' + \lambda'' t'')} \tag{9.4}$$

Similar considerations apply to multicomponent systems.

The exponential density function, like other statistical density functions, also has a characteristic value called the *mean*. This is obtained for all distributions by forming what is called the *first moment* $t \cdot f(t)$ of the density function $f(t)$ and integrating the first moment over the entire range of $f(t)$. If we make this operation for the exponential density function of (9.1), we obtain the mean of the exponential function, which we call the *mean time between failures:*

$$m = \int_0^\infty t f(t) \, dt = \int_0^\infty t \lambda e^{-\lambda t} \, dt = \frac{1}{\lambda} \tag{9.5}$$

Thus, only in the exponential case, the mean time between failures (the mean) m is equal to the reciprocal value of the failure rate λ.

Other, nonexponential density functions also have a mean value—for instance, the mean life M of a normal distribution. In these cases, however, the mean value is not the reciprocal of the failure rate. Although M is a constant in the normal distribution, its failure rate is not constant but is a variable function of the time, as we have seen in Chapter 6, Figure 6.5.

The mean value can generally also be obtained by integrating the reliability function $R(t)$ over its entire range, so that m is then numerically equal to the area under the reliability curve from $t = 0$ to $t = \infty$.* In the exponential case we then obtain

$$m = \int_0^\infty R(t)\, dt = \int_0^\infty e^{-\lambda t}\, dt = - \left[\frac{1}{\lambda} e^{-\lambda t} \right]_0^\infty = \frac{1}{\lambda} \qquad (9.6)$$

which is identical with (9.5).

The proof of the identity is obtained as follows:

$$m = \int_0^\infty t f(t)\, dt = \int_0^\infty t \left(- \frac{dR}{dt} \right) dt$$

which, integrated by parts, yields

$$m = - \left[tR \right]_0^\infty + \int_0^\infty R\, dt$$

For all practical purposes, the first term, $-[tR]_0^\infty$, is zero, since

$$R = \exp \left(- \int_0^t \lambda\, dt \right)$$

and

$$tR = \frac{t}{\exp \left(+ \int_0^t \lambda\, dt \right)}$$

which in the limit must become zero as t increases to infinity, if, after an arbitrarily long time, the failure rate λ either remains at some constant value greater than zero or increases steadily.

Having proved that in the exponential case $\lambda = 1/m$, we can rewrite Equations (9.1), (9.2), and (9.3) by substituting $1/m$ for λ:

$$f(t) = \lambda \cdot e^{-\lambda t} = \frac{1}{m} e^{-t/m} \qquad (9.7)$$

$$Q(t) = 1 - e^{-t/m} \qquad (9.8)$$

$$R(t) = e^{-t/m} \qquad (9.9)$$

We shall make frequent use of both notations, with m and with λ, in further discussions.

Exponential reliabilities are easy to multiply by simply adding the exponents. For instance, if one component has the reliability

$$R_1(t) = e^{-\lambda_1 t} = \exp(-t/m_1)$$

---

* Boeing Airplane Company's *Reliability Handbook*, Document D6-2770 (September 1957), prepared by Igor Bazovsky, Part I, p. 6.

and another component has the reliability

$$R_2(t) = e^{-\lambda_2 t} = \exp\left(-t/m_2\right)$$

the product of the two reliabilities is

$$R(t) = R_1(t) \cdot R_2(t) = e^{-\lambda_1 t - \lambda_2 t} = e^{-(\lambda_1 + \lambda_2)t}$$

$$= \exp\left[-t\left(\frac{1}{m_1} + \frac{1}{m_2}\right)\right] \tag{9.10}$$

This product reliability is, then, the probability that neither component will fail in the time t, or, which amounts to the same, the probability that both components will survive in the time t. The same multiplication rule can be applied to any number of independently failing components if we want to know the probability that none of them will fail in a given operating time t. This rule can also be applied to two different time periods, as was shown in Equation (9.4). The reliability $R(t)$ in (9.4) then amounts to the probability that the component will not fail in either of the two time periods, or, that it will survive through both time periods t' and t''. If the component's failure rate remains the same in both time periods, then its reliability for the whole operating period $t = t' + t''$ will be

$$R(t) = e^{-\lambda(t'+t'')} = e^{-\lambda t} \tag{9.11}$$

Situations such as that described by Equation (9.4), i.e., when components and systems change failure rates in different time periods because the stress level of the environment has changed, occur in all airborne systems. For instance, in aircrafts the failure rates of some components will be many times higher during the take-off than while cruising, because the stress level changes very significantly.

A similar situation applies to missiles and space vehicles. During the firing, and as long as the propulsion system is in operation, the failure rates will be much higher than after the motors or rockets are cut off. Mechanical shocks, vibration, and thermal shocks are mainly responsible for this phenomenon.

These changes of the failure rate with the stress level are an entirely different phenomenon from that discussed in the case of early failures and wearout. In the case of early failures, the failure rate is decreasing due to the elimination of substandard specimens, or, in the case of a single component, its instantaneous probability of failure is decreasing as it becomes assured that the component is not substandard; in the case of wearout failures, the failure rate is increasing with a gradual degradation of the component's strength. In the case of failure rate changes with changing stress level, we assume that the component has reached a constant strength which is not changing, so that it behaves exponentially and its failure rate changes only when the stress level changes, but at the new stress level it again behaves exponentially. If the stress level returns to its former value, the failure rate of the component will also return to its former value, if the component has maintained its original strength. That means that the component remains basically exponential, whereas this is not the case with early and wearout failures.

In Chapter 4 we derived the exponential function from the basic notion of the probability concept. This function is remarkably well suited to describe events which occur in the time domain, where it is diffi-cult to enumerate the number of occurrences and the number of non-occurrences of the event. For instance, during a thunderstorm we may observe ten lightning flashes, but we cannot say how many times there were no lightning flashes during that storm.

To calculate the probability of an event A, we usually require infor-mation about how often the event has occurred and how often its com-plementary event B, which designates the nonoccurrence of the event A, has occurred. Knowing these two numbers in a trial, we then have $P(A) + P(B) = 1$. For instance, if we draw a sample of n components from a large lot in which there are 90 per cent good and 10 per cent bad components, we know that the probability of drawing a good one in a single draw is $p = 0.9$ and the probability of drawing a bad one is 0.1, so that $p + q = 0.9 + 0.1 = 1$. If we want to know the probability dis-tribution in n draws, we obtain this from the binomial distribution as

$$(p + q)^n = p^n + n \cdot p^{n-1} \cdot q + \frac{n(n-1)}{2!} \cdot p^{n-2} \cdot q^2 + \cdots + q^n = 1 \quad (9.12)$$

where each term has a definite probability meaning. So if our trial with the components consists of $n = 10$ draws, we obtain a probability of $p^n = 0.9^{10}$ of drawing 10 good components, a probability of

$$n \cdot p^{n-1} \cdot q = 10 \cdot (0.9)^9 \cdot (0.1)$$

of drawing 9 good and 1 bad component in a trial, etc., and finally a probability of $q^n = 0.1^{10}$ of drawing 10 bad components. The sum total of all these probabilities must equal unity because we have exhausted all possibilities in which these 10 draws can end. We now know the definite probabilities for any combination of good and bad components in a single trial in which we draw 10 or any other number of components.

However, the binomial distribution cannot be directly applied to the calculation of event probabilities in the time domain because the total number n of the favorable and unfavorable events is not easily computed; in fact, it is usually unknown. As we said before, we may observe ten flashes of lightnings in a storm, but to calculate the probability of the flashes we would have to know how often the complementary event B, the nonoccurrence of lightning, has occurred. So here we are left with only the binomial distribution.

The Poisson distribution now helps to overcome the difficulty. It uses the number e, the base of the natural logarithm, to obtain an equivalent expression for $(p + q)^n = 1$. Poisson obtains this equivalent expression from

$$e^{-x} \cdot e^x = 1 \qquad (9.13)$$

where e^x is expanded into an infinite series

$$e^x = 1 + x + \frac{x^2}{2!} + \frac{x^3}{3!} + \frac{x^4}{4!} + \cdots$$

so that we can write for (9.13)

$$e^{-x}\left(1 + x + \frac{x^2}{2!} + \frac{x^3}{3!} + \frac{x^4}{4!} + \cdots\right) = 1 \qquad (9.14)$$

When we take x to be the expected or average number of occurrences of an event A, the terms in Equation (9.14) contain definite probability meanings. For instance, e^{-x} is the probability that the event will not be observed, xe^{-x} is the probability that the event will be observed once, $(x^2/2!) \cdot (e^{-x})$ is the probability that the event will be observed twice, etc. Therefore, all we need to know is x, the average number of occurrences of an event A, and we can immediately calculate all the probabilities of the various, possible occurrences of that event without having a knowledge of n.*

When components are operated in the time domain, we have to refer the average number of the occurrences of an event connected with the component's operation, for instance, failure, to some length of time. Taking $x = \lambda t$ as the average number of failures in time t, so that λ is the average number of failures in a time unit (hour), we obtain the probability that no failure will occur in the time t, which we call reliability,

$$R = e^{-\lambda t} \qquad (9.15)$$

For the probability that exactly one failure will occur in the same period t, we obtain

$$Q_1 = (\lambda t)e^{-\lambda t} \qquad (9.16)$$

and for the probability that exactly two failures will occur in the same period t,

$$Q_2 = \frac{\lambda^2 t^2}{2!} e^{-\lambda t} \qquad (9.17)$$

etc.

* M. J. Moroney, *Facts from Figures*, Penguin Books Inc., Baltimore, 1954, pp. 96–98.

We can simplify our calculations by adding up all Q's, i.e., $Q = Q_1 + Q_2 + Q_3 + \cdots$ and calling this sum the *unreliability*, which is the probability that one or more failures will occur in time t, or simply the probability that there will be at least one failure in time t. This results in $R + Q = 1$ or

$$Q = 1 - R = 1 - e^{-\lambda t} \tag{9.18}$$

Therefore, all we need to know is the average number of failures per unit time and we can calculate the probability of no failure, or reliability, as well as the probability of failure, or unreliability, for any length of time t. This is a most remarkable fact of the Poisson distribution which so frequently occurs in nature, and which is based on the number $e = 2.71829\ldots$.

The Poisson distribution can also be derived directly from the binomial distribution as a limit case.*

The computation of the unreliability Q of a component will often facilitate reliability calculations of complex systems, as we shall see in the next chapters—especially where parallel operation of components is involved. In certain cases of redundant component arrangements we shall also make use of the calculation of the partial unreliabilities which we denoted Q_1, Q_2, Q_3, etc., and therefore, of the probabilities that exactly one, two, three, etc., failures occur. We shall need this when we discuss the so-called stand-by operations.

---

* A. Hald, *Statistical Theory with Engineering Applications*, John Wiley & Sons, Inc., New York, 1952, chap. 21, par. 21.7 ("The Binomial Distribution and the Poisson Distribution").

Chapter 10

RELIABILITY OF
SERIES SYSTEMS

IN THE PRECEDING CHAPTERS we learned that reliability is the probability of survival and that reliability can be expressed mathematically throughout the entire life of a component. Also, we have learned that when a component is operated within that period of its life which we call the *useful life*, it will display a constant failure rate λ, and its reliability is then expressed by a simple exponential function $e^{-\lambda t}$. We have further seen that this simple function also applies to situations where components are mixed by age and are replaced only as they fail regardless of the cause of the failures.

But reliability is not confined to single components. What we really want to evaluate is the reliabilities of systems, simple as well as extremely complex, and to use these evaluation techniques for designing reliable systems or for gaining reasonable assurance in advance that a design will meet certain reliability requirements.

System reliabilities are calculated by means of the calculus of probability. To apply this calculus to systems, we must have some knowledge (given or otherwise obtainable) of the probabilities of its parts, since they affect the reliability of the system. Therefore, to calculate system reliability we must have a knowledge about the reliabilities of those components which can cause the system to fail.

Component reliabilities are derived from tests which yield information about failure rates. When a new component is designed and built, no measure of electrical, mechanical, chemical, or structural calculations can tell us the failure rate of that component.

We can foretell what the electrical, thermal, or mechanical strength of a newly designed component ought to be and we also know that the stronger a component is, the smaller its failure rate—but the actual value of this failure rate can be obtained only by means of statistical procedures such as reliability tests, because of two main factors which govern the probability of survival of a component: (1) the uncertainties of the production process, and (2) the uncertainties of the stresses which a component must withstand in operation. The failure rate is the result of the interaction of the strength designed and built into a component with the operational stress spectrum. In reliability tests we actually measure the failure rate of a component, which means we measure its instantaneous probability of failure at a given set of environmental and operating stress conditions.

Thus, system reliability calculations are based on two important operations: (1) as precise as possible a measurement of the reliability of the components used in the system environment, and (2) the calculation of the reliability of some complex combination of these components.

We know already that more extensive testing will yield better information about the true reliability of the components. The calculation of the system reliability is then carried out by means of a few comparatively simple and very exact probability theorems. Although it often happens that the set of conditions under which a component operates in a system differs from the test conditions from which we obtained the failure rate, there are methods (to be discussed later) which allow us to estimate the effects of the system environment on the component failure rate, or, we can test a component for reliability in the system itself or in a simulated system. But once we have the right figures for the reliabilities of the components in a system, or good estimates of these figures, we can then perform very exact calculations of system reliabilities even when the systems are the most complex combinations of components conceivable. The exactness of our results does not hinge on the probability calculations because these are perfectly accurate; rather, it hinges on the exactness of the reliability data of the components. Later we shall see how we can make component reliability tests exact enough for our purposes without going into "infinitely" large samples.

In system reliability calculations we shall at first use the following basic rules of the probability calculus:

1. If A and B are two independent events with the probabilities $P(A)$ and $P(B)$, then the probability that both events will occur is the product

$$P(AB) = P(A) \cdot P(B) \qquad (10.1)$$

2. If the two events can occur simultaneously, the probability that either A or B or both A and B will occur is

$$P(A \vee B) = P(A) + P(B) - P(A) \cdot P(B) \qquad (10.2)$$

3. If the two events are mutually exclusive so that when one occurs the other cannot occur, Equation (10.2) simplifies to

$$P(A \lor B) = P(A) + P(B) \qquad (10.3)$$

4. If the two events are complementary in addition to being mutually exclusive, that is, if A does not occur B must occur and vice versa, we obtain from (10.3)

$$P(A) + P(B) = 1 \qquad (10.4)$$

Translated into the language of reliability where component survival and failure are two complementary and mutually exclusive events so that $R + Q = 1$ for the same period of operation, and where we count it as a failure only if a component fails independently of the failure of another component, we obtain the following basic formulas for the calculation of reliability of a combination of two or more components:

1. If a component has a reliability R_1 and another component has a reliability R_2, then the probability that both components will survive an operating time t is

$$R_s(t) = R_1(t) \cdot R_2(t) = \exp\left[-\int_0^t \lambda_1 \, dt\right] \cdot \exp\left[-\int_0^t \lambda_2 \, dt\right] \qquad (10.5)$$

where λ_1 and λ_2 are the failure rates of the components and can be time variable or constant.

2. The probability that either one or both of the components will fail is

$$Q_s(t) = Q_1(t) + Q_2(t) - Q_1(t) \cdot Q_2(t)$$
$$= 1 - R_1(t) + 1 - R_2(t) - [1 - R_1(t)] \cdot [1 - R_2(t)]$$
$$= 1 - R_1(t) \cdot R_2(t) = 1 - R_s(t) \qquad (10.6)$$

3. The probability that either one or both of the components will survive is

$$R_p(t) = R_1(t) + R_2(t) - R_1(t) \cdot R_2(t)$$
$$= \exp\left[-\int_0^t \lambda_1 \, dt\right] + \exp\left[-\int_0^t \lambda_2 \, dt\right]$$
$$- \exp\left[-\int_0^t (\lambda_1 + \lambda_2) \, dt\right] \qquad (10.7)$$

4. The probability that both components will fail is

$$Q_p(t) = Q_1(t) \cdot Q_2(t) = [1 - R_1(t)] \cdot [1 - R_2(t)]$$
$$= 1 - R_1(t) - R_2(t) + R_1(t) \cdot R_2(t) = 1 - R_p(t) \qquad (10.8)$$

Cases 1 and 2 are complementary events so that $R_s + Q_s = 1$, because the complementary event to both components surviving is the event when not both of them survive, and that event comprises three possibilities: either one component fails, or the other component fails, or both fail. We therefore call R_s and Q_s the *reliability and the unreliability of a series connection of components or a series system of components*, which means that if any one component fails, the system has failed. We shall see that all complex systems belong in this category.

Cases 3 and 4 are also complementary events so that $R_p + Q_p = 1$, because the complementary event to both components failing is the event when either or both components survive. We call R_p and Q_p the *reliability and the unreliability of a parallel connection of components or a parallel redundant system*, which means that if one component fails, there is another component operating in parallel which will continue to perform the required function and, therefore, such a system of two parallel components does not fail if one component fails. Obviously, if there are more than two components operating in parallel—for instance, n components— the system will not fail as long as at least one of these components remains operative. Therefore, out of n parallel components, normally $n - 1$ components can fail without causing the system to fail, provided that the single surviving component is sufficient to perform the necessary function.

The equations above are valid for exponential as well as for non-exponential components. When nonexponential components are considered, we must take into account that their failure rates are not constant but are a function of the age T of these components. Therefore, as explained in previous chapters, for an operating period t for which the reliability is to be computed, component failure rates corresponding to the component age at the given operating period must be used.

Fortunately, in most cases we can do very well by assuming that components behave exponentially in systems when the system time is taken as the basis for observing component failures or when components operate only within their useful life period. In the exponential case, when the failure rates are constant, Equations (10.5) through (10.8) simplify to

$$R_s(t) = e^{-\lambda_1 t} \cdot e^{-\lambda_2 t} = e^{-(\lambda_1 + \lambda_2)t} \tag{10.9}$$

$$Q_s(t) = 1 - e^{-(\lambda_1 + \lambda_2)t} \tag{10.10}$$

$$R_p(t) = e^{-\lambda_1 t} + e^{-\lambda_2 t} - e^{-(\lambda_1 + \lambda_2)t} \tag{10.11}$$

$$Q_p(t) = (1 - e^{-\lambda_1 t})(1 - e^{-\lambda_2 t}) \tag{10.12}$$

Complete systems usually consist of a large number of components or units combined in series, which means that if any of these components or units fails, the system fails. In some cases certain less reliable components in a system are backed up by parallel components to increase system reliability by parallel redundancy, and sometimes even a whole group of components is backed up by an equal or similar group, operating in parallel with the first group. Such parallel arrangements of two or more components or groups can be considered as a single unit in series within the system so that if the unit as a whole fails, the system fails. In this way we again arrive at a series arrangement of reliabilities, or a series system. Therefore, the reliability of every complex system can be expressed as the product of the reliabilities of all those components and units on whose satisfactory operation the system depends for its survival. For n components or units in series the system reliability is given by

$$R_s = R_1 \cdot R_2 \cdot R_3 \cdot R_4 \cdot \ \cdots \ \cdot R_n = \prod_{i=1}^{n} R_i \qquad (10.13)$$

where R_i is the reliability of the ith component or unit in series in the system. Here R_i can be exponential or nonexponential. If it represents a single component it will normally be exponential; but if it represents a unit with a redundant arrangement of components, R_i will not be exponential because the failure rate of such a unit is not constant anymore, as we shall see later. Equation (6.13) still applies to all situations and is therefore the fundamental equation for complex system reliability. It is called the *product law of reliabilities.*

When all R_i's in the system are exponential, Equation (6.13) for the system's reliability becomes simple to evaluate:

$$R_s = e^{-\lambda_1 t} \cdot e^{-\lambda_2 t} \cdot e^{-\lambda_3 t} \cdot \ \cdots \ \cdot e^{-\lambda_n t}$$

$$= e^{-(\lambda_1 + \lambda_2 + \lambda_3 + \cdots + \lambda_n)t} = \exp\left(-\sum_{i=1}^{n} \lambda_i t\right) \qquad (10.14)$$

Thus, all we need do is add up the constant failure rates of all the series components (in the system time scale, and at the system stress level), multiply this sum with the operating period t, and obtain the value of R_s from exponential tables.

As a simple example of a preliminary reliability analysis, let us consider an electronic circuit consisting of 4 silicon transistors, 10 silicon diodes, 20 composition resistors, and 10 ceramic capacitors in continuous series operation and assume that the wiring (printed circuit) and the solder connections are 100 per cent reliable. Further, let us assume that the components in the assembled and packaged circuit will operate at their rated values of voltage, current, and temperature (for instance, let the temperature inside the package be 85°C) and that under these stress conditions the components have the following failure rates:

Silicon diodes:	λ_d = 0.000002 each
Silicon transistors:	λ_t = 0.00001 each
Composition resistors:	λ_r = 0.000001 each
Ceramic capacitors:	λ_c = 0.000002 each

To estimate the reliability of this circuit we first add up all failure rates:

$$\Sigma \lambda_i = 10\lambda_d + 4\lambda_t + 20\lambda_r + 10\lambda_c = 0.0001$$

This sum is the expected hourly failure rate λ_s of the whole circuit. Fig-

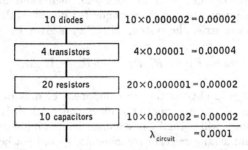

Fig. 10.1. Simplified reliability block diagram of a series circuit.

ure 10.1 shows its reliability block diagram. The estimated reliability of the circuit is then

$$R_s(t) = e^{-0.0001 \cdot t}$$

for an operating time t. For a 10-hour operation the reliability or probability that the circuit will not fail in 10 hours is

$$R_s(10) = e^{-0.0001 \cdot 10} = e^{-0.001} = 0.999 = 99.9\%$$

Thus, this circuit could be expected to operate on the average without failure in 999 such 10-hour operations and to fail only once in 1000 operations. Therefore, its unreliability is $0.001 = 1 - 0.999$.

As shown in the above calculation, the expected failure rate of the circuit is 0.0001 per hour. Therefore, its expected mean time between failures is

$$m = \frac{1}{\lambda} = \frac{1}{0.0001} = 10,000 \text{ hours}$$

This does not mean that the circuit could be expected to operate without failure for 10,000 hours. We know already from the exponential function that its chance to survive for 10,000 hours is only about 37 per cent, and its chance to fail in 10,000 hours is therefore 63 per cent. But the knowledge of the mean time between failures is important since it is a parameter which can be measured and which completely defines the reliability of a series system during its useful life.

The precision of the calculated value of the circuit's reliability depends on the accuracy of its components' failure rate figures. Those used in the example above are only assumed figures and not actual failure rate values; these must be obtained either from the manufacturers of the components, from reasonably extensive observation in service, or by testing.

The estimated mean time between failures of 10,000 hours for the analyzed circuit may fall short of the value required for a specific application. The requirement may be for a mean time between failures of 100,000 hours. What can the designer do to upgrade the reliability of the circuit so as to meet the requirement? The answer lies in derating the components.

Component failure rate figures apply to definite operating stress conditions—for instance, to an operation at rated voltage, current, temperature, and at a predicted level of mechanical stresses, such as shock and vibration. Failure rates usually change radically with changes in the stress levels. If a capacitor is operated at only half of its rated voltage, its failure rate may drop to one-thirtieth of the failure rate at full rated voltage operation. A similar sensitivity applies to temperature changes. The failure rates of most electronic components are strongly affected by the voltage, current, and temperature levels at which the components are applied.

Thus, to upgrade the reliability of the circuit it becomes necessary to reduce the stresses acting on the components—that is, to use components of higher voltage and current ratings, and to make provisions for a reduction of the operating temperature levels. Using these techniques, component failure rate reductions by a factor of ten are often easily achieved. Some component manufacturers supply failure rate derating curves from which the failure rates of their components can be directly read for various stress levels. We proceed then by performing first a stress analysis of the circuit, determining for each component the operating voltage or wattage as a percentage of its rated value, and estimating (and later measuring as well) the operating temperature inside the circuit package. Then, from the failure rate derating curves, we derive for each individual component its failure rate under the evaluated stress conditions, add up these failure rates, and add to the sum another ten per cent to take care of connections, wiring, etc. When mechanical stresses (shock, vibration) are also involved, these have to be considered at the stress analysis stage. Their effect on electronic component failure rates depends on the packaging techniques used for the circuit modules. Shock-resistant packaging and elimination of resonant vibrations will greatly help to suppress such effects.

The above discussion makes it quite obvious that the architect of an equipment's reliability is the designer himself. He can achieve large gains in reliability by operating the components at low stress levels. Also, whenever possible, he should reduce the number of components used. Thus, when designing the circuits and their packaging, he should always keep two things in mind:

1. Do not overstress the components, but operate them well below their rated values, including temperature. Provide good packaging against shock and vibration, but remember that in tightly packaged equipment without adequate heat sinks, extremely high operating temperatures may develop which can kill all reliability efforts.
2. Design every equipment with as few components as possible. Such simplification of the design increases reliability and makes assembly and maintenance easier.

Very little can be done about reliability once an equipment has been designed and goes into production. Sometimes components of higher ratings can be substituted if space in the package permits their use. But, in general, if an equipment turns out to be unreliable because insufficient attention was given to reliability during the design stage, only extensive redesign can help, and money and time are lost. A good designer, therefore, will have a fine appreciation for and an adequate knowledge of reliability.

As already mentioned in Chapter 2, the failure rates of some components such as switching devices are more meaningfully expressed in terms of "failures per operating cycle" than in terms of "failures per hour of component operation." When we have such components in a system, we must convert their failure rates from per-operating-cycle units to per-hour-of-system-operation units before Equation (10.14) can be numerically evaluated. A conversion of failure rates to per-hour-of-system-operation also becomes necessary for components which operate only sporadically during the system's operation, although their failure rates are given in time units such as failures per hour of component operation.

The time t in Equation (10.14) is the system operating time. Only when a component operates continuously in the system will the component's operating time be equal to the system's operating time. For instance, if a component has to operate for only one-tenth of the time the system is operating, i.e., for $t/10$ hours for every t hours of system operation, and when we assume that in a nonoperating condition the component has zero failure rate, the component's reliability for t hours of system operation is

$$\exp\left(-\frac{\lambda' t}{10}\right) = \exp\left[-\left(\frac{\lambda'}{10}\right)t\right]$$

where λ' is the component's failure rate per one hour of component operation. Thus, in the system the component appears to assume a failure rate of $\lambda'/10$. And if the component were to operate in the system only $t/1000$ hours for every t hours of system operation, its failure rate in terms of the system time scale would be $\lambda'/1000$. Or, in general, when a component operates on the average for t_1 hours in t system operating hours, it assumes in the system's time scale a failure rate of

$$\lambda = \frac{\lambda' t_1}{t} \tag{10.15}$$

We would then use this converted component failure rate in Equation (10.14) for components which operate only sporadically, because this is the failure rate which such components assume in the system's time scale. If t_1 is a very small fraction of t, a component can appear to be extremely reliable in terms of the system operating time t, even if it had a comparatively high failure rate when operated continuously.

Equation (10.15) is based on the assumption that in a nonoperating or de-energized condition the component has a zero failure rate even though the system is in operation. This is not always the case. Components may exhibit some failure rates even in their quiescent or idle condition while the system is operating.

If the component has a failure rate of λ' when operating and λ'' when de-energized, and it operates for t_1 hours every t hours of system operation, so that it is de-energized for $t_2 = t - t_1$ hours, the system will see the component behaving with an average failure rate of

$$\lambda = \frac{\lambda' t_1 + \lambda'' t_2}{t} \tag{10.16}$$

per system operating hour. Equation (10.16) is thus the failure rate of such a component in the system time scale.

If the failure rate of a component is expressed in terms of operating cycles, i.e., as λ_c per one operating cycle, and if the component performs on the average c operations in t system hours, the system will see the component behave with a failure rate of

$$\lambda = \frac{c \lambda_c}{t} \tag{10.17}$$

But if this component also has a time dependent failure rate of λ' while energized, i.e., after it has been switched on, and a failure rate of λ'' when de-energized and the system is still operating, the component assumes in the system time scale a failure rate of

$$\lambda = \frac{c \lambda_c + t_1 \lambda' + t_2 \lambda''}{t} \tag{10.18}$$

per system operating hour.* Obviously, in this equation $t_1 + t_2 = t$, which is the system's mission time for which the system's reliability is to be calculated.

Equation (10.18) can be considered as a general equation for the evaluation of component failure rates in terms of system time scale. For most components λ'' can be neglected, except in cases where the stresses acting on a component in an operating system are very high even though the component itself is not energized. As to λ_c and λ', we find that the failure rate of some devices—in particular, switching devices— is almost exclusively governed by λ_c, so that λ' can also be neglected; other components—in particular, continuously energized components— have primarily a time dependent failure rate λ'. However, there is a category of components which are not switching devices and where both λ_c and λ' must be considered. These are the components which are essentially continuously energized devices but in which extremely high temperature gradients develop when they are being switched on or off. Such is the case with incandescent lamps such as signal lamps, etc. It is a well known phenomenon that lamps usually burn out a few instants after they are switched on, although on other occasions they can also fail in continuous operation. Thus lamps definitely have a λ_c and a λ' failure rate. The same is true of electron tubes.

When components which exhibit a λ_c failure rate are part of a system, system reliability as given by Equation (10.14) then obviously depends on the frequency at which these components are switched. If we use the general expression (10.18) for the failure rate of exponential components, the reliability of a series system of n components becomes

$$R = \exp\left(-\sum_{}^{n} \lambda t\right) = \exp\left[-\sum_{}^{n} (c_i \lambda_{ci} + t_{1i}\lambda_i' + t_{2i}\lambda_i'')\right] \quad (10.19)$$

where c_i is the number of switching cycles of the ith component in $t = t_{1i} + t_{2i}$ system operating hours.

If the ith component is, for instance, a relay which performs only one switching cycle in t hours, i.e., it turns the system on, stays on for t hours, and turns the system off after t hours so that $c_i = 1$, and if we assume that λ' and λ'' of the relay are zero, the relay's contribution to the system's failure rate is just λ_c and the relay's reliability for this operation of t hours is $\exp(-\lambda_c)$, which is independent of time. If we now assume that all other components in the system are 100 per cent reliable, system reliability would be $R = \exp(-\lambda_c)$ for one mission, regardless of how long the mission might last, if during that mission the relay is required to perform only one operating cycle, i.e., to turn the system on and turn it

---

* The component's mean time between failures in terms of the system operating time t is then the reciprocal of λ.

off again. Let us say the mission is of 100 hours duration. Now let us consider what happens with the reliability of the same system when, in another application, the system is operated intermittently so that it is 10 times switched on and off in 100 hours. Obviously, the reliability of the system for each individual operation is again $\exp(-\lambda_c)$, but for 10 operations in 100 hours it becomes

$$R = [\exp(-\lambda_c)]^{10} = \exp(-10\lambda_c)$$

Thus, for the same 100 hours, system reliability is now one order of magnitude smaller.

In actual cases the effect of switching will not be so pronounced because the sum of the time dependent failure rates of the other components in a system is usually much larger than the per-cycle failure rate of a switching device which performs only one switching cycle during a mission. However, if a switching device performs a large number of cycles during a mission, or if a system contains a large proportion of switching devices or of devices which are sensitive to high temperature gradients caused by switching (e.g., lamps, tubes), the switching frequency of these devices and of the system must be considered in system numerical reliability analysis.*

The question of whether or not it is better and more economical to leave such systems "on" even during intervals when system operation is not required sometimes comes up. Thus, is it preferable to switch the system off and turn it on again only when needed, or leave it on all the time?

Which procedure is better from the reliability standpoint depends on the ratio of the system's probability of surviving T hours of unneeded operation to its probability of surviving an off-on switching cycle. We obtain a criterion for the choice of procedure by forming the ratio of the expected number of system failures for these two probabilities:

$$p = \frac{\sum\limits^{n} (c_i\lambda_{ci} + T_{1i}\lambda_i' + T_{2i}\lambda_i'')}{\sum\limits^{n} \lambda_{ci}} \tag{10.20}$$

where $T_1 + T_2 = T$. If $p > 1$, higher reliability is achieved by turning the system off for the T-hour interval when system operation is not needed. If $p < 1$, higher reliability is achieved by leaving the system on for the T-hour interval so that the system is already on when the next operating mission begins. Obviously, the ratio p can become smaller than unity only if the numerator is smaller than the denominator. This can

---

* See "Effects of Cycling on Reliability of Electronic Tubes and Equipment," ARINC (Aeronautical Radio, Inc.) Publication No. 101-36-160 (June 30, 1960).

happen, for instance, if during the interval T when the system is left on, no components perform any switching functions so that $c = 0$. If during the T hours all system components are continuously energized, the expected number of system failures is $T \sum\limits^{n} \lambda_i$. This must be smaller than the expected number of system failures per one off-on switching operation, i.e., $\sum\limits^{n} \lambda_{ci}$. The requirement for p to be smaller than unity can thus be fulfilled only if the length of the interval T is smaller than the ratio of the system's failure rate per one switching cycle to the system's failure rate per one hour of continuous operation:

$$T < \frac{\sum\limits^{n} \lambda_{ci}}{\sum\limits^{n} \lambda_i} \qquad (10.21)$$

It is obvious that T can be significant only for systems which contain components with comparatively high λ_c failure rates and when these components are not subject to switching during the normal mode of system operation.

As to the economy of leaving a system switched on when not needed, the cost of consumed energy during unnecessary operation is important. But even more important is that during an energized condition most components are subject to a wearout process or to a gradual degradation of their tolerance limits, whereas these phenomena are essentially halted in a de-energized state.

Chapter 11

RELIABILITY OF
PARALLEL SYSTEMS

W HEN A SYSTEM must be designed to a quantitatively specified reliability figure, it is not enough that the designer simply reduce the
number of components and the stresses acting on them. He must, during
the various stages of the design, perform reliability calculations to be sure
that his design is proceeding so that it will reach the required quantitative reliability goal. If very high system reliabilities are required, the
designer must duplicate components, and sometimes whole circuits, to
fulfill such requirements. In other words, he must make use of parallel
reliabilities, called *parallel redundancy*.

The calculation of parallel reliabilities is based on Equations (10.7)
and (10.8), which give the reliability of a combination of two components
operating in parallel. In the exponential case the equations were reduced
to (10.11) and (10.12).

But there are cases where more than two components operate in
parallel. Then we have to extend the probability theorem 2, given by
Equation (10.2), to n events which can occur simultaneously. Although
we assume that the reader has a knowledge of the theory of probability,
we give the rule for the calculation of the probability that of three events
A, B, and C, having the probabilities $P(A)$, $P(B)$, and $P(C)$, either A,
or B, or C, or any combination of these three will occur:

$$P(A \lor B \lor C) = P(A) + P(B) + P(C) - P(A)P(B)$$
$$- P(A)P(C) - P(B)P(C) + P(A)P(B)P(C) \quad (11.1)$$

If the three events have the same probabilities, i.e., if $P(A) = P(B) = P(C) = P$, we then obtain

$$P(A{\vee}B{\vee}C) = 3P - 3P^2 + P^3 \qquad (11.2)$$

Similar probability equations exist for four or more events.

Applying Equation (11.1) to the case of three components operating in parallel, we calculate the reliability of such parallel arrangement as the probability that any one of the three components will survive or any combination of them will survive, which amounts to the probability that at least one will survive. When the three components have the failure rates λ_1, λ_2, λ_3, this reliability will be

$$R_p(t) = e^{-\lambda_1 t} + e^{-\lambda_2 t} + e^{-\lambda_3 t} - e^{-(\lambda_1+\lambda_2)t}$$
$$- e^{-(\lambda_1+\lambda_3)t} - e^{-(\lambda_2+\lambda_3)t} + e^{-(\lambda_1+\lambda_2+\lambda_3)t} \qquad (11.3)$$

And when the three components have equal failure rates of λ each,

$$R_p(t) = 3e^{-\lambda t} - 3e^{-2\lambda t} + e^{-3\lambda t} \qquad (11.4)$$

Similarly, we could develop the reliability equations for four components in parallel, etc., but this is a rather tedious process. There is a much simpler way to calculate parallel reliabilities. We make use of the equality $R + Q = 1$, calculate first the unreliability Q, and by deducting this from 1 we obtain the reliability.

From Equations (10.8) and (10.12) we know that the probability of two components failing is $Q = Q_1 \cdot Q_2$. It follows that the probability of three components failing, or the unreliability of an arrangement of three components in parallel, is $Q = Q_1 \cdot Q_2 \cdot Q_3$ and the unreliability of n components in parallel is

$$Q_p(t) = Q_1(t) \cdot Q_2(t) \cdot Q_3(t) \cdot \ \cdots \ \cdot Q_n(t) = \prod_{i=1}^{n} Q_i(t) \qquad (11.5)$$

This is called the *product law of unreliabilities in parallel operation.* The reliability of n components in parallel is then

$$R_p(t) = 1 - Q_p(t) = 1 - \prod_{i=1}^{n} Q_i(t) \qquad (11.6)$$

The procedure of first computing the unreliability Q_p as the product of the unreliabilities of the parallel components greatly facilitates the numerical calculations of the reliability R_p of parallel arrangements.

In practice, a simple approximating procedure is often used for a quick estimation of component unreliabilities. When $e^{-\lambda t}$ is developed into an infinite series, we obtain

$$R = e^{-\lambda t} = 1 - \lambda t + \frac{\lambda^2 t^2}{2!} - \frac{\lambda^3 t^3}{3!} + - + - \cdots \qquad (11.7)$$

and

$$Q = 1 - e^{-\lambda t} = \lambda t - \frac{\lambda^2 t^2}{2!} + \frac{\lambda^3 t^3}{3!} - + - + \cdots \qquad (11.8)$$

When the exponent λt is much smaller than one, $\lambda^2 t^2/2!$ and the following terms of the series can be neglected with good approximation:

$$Q = 1 - e^{-\lambda t} \cong \lambda t \qquad (11.9)$$

$$R = e^{-\lambda t} \cong 1 - \lambda t \qquad (11.10)$$

These approximations can be used in rough estimating work when $\lambda t < 0.1$. However, when more complex calculations are involved which call for high precision, tables giving the values of e^{-x} with at least five decimals, and possibly more, should be used. To form the product of several small numbers each of which has five or more decimals, the use of reliable calculating machines is recommended. In fact, exact reliability calculations require the use of such machines all the time.

Components operating in parallel are very often equal. Equations (11.5) and (11.6) then simplify to

$$Q_p = Q^n \qquad (11.11)$$

$$R_p = 1 - Q^n \qquad (11.12)$$

where Q is the unreliability of one component.

All the equations which we derived for the parallel operation of components apply also to the parallel operation of groups of components and parallel operating circuits.

When redundancy becomes a necessity, it is often preferable, for economic design reasons, to duplicate a group of components or a whole circuit instead of duplicating each component separately.* In such cases we first calculate the unreliability Q_i of the individual series groups or circuits and we then proceed with the calculation of the unreliability Q_p and the reliability $R_p = 1 - Q_p$ of the whole parallel arrangement. Thus we deal with a series group of components or with a series circuit as if it were a single component being duplicated.

We have seen that a series system consisting of exponential components has a constant failure rate λ_s which is the sum of all the component failure rates, and because such a system is exponential its mean time between failures is $m_s = 1/\lambda_s$. However, this does not apply to parallel systems. The instantaneous failure rate of a parallel system is a variable function of the operating time t, although the mean time between fail-

---

* Higher reliabilities may be obtained by separate duplication of components, as shown at the end of this chapter.

ures m_p is still a constant and can be computed from the system's reliability R_p by means of Equation (9.6).*

When two components with the failure rates λ_1 and λ_2 operate in parallel we illustrate such an arrangement by the block diagram in Figure 11.1. The reliability R_p of this parallel system is given by Equation (10.11) as

$$R_p = e^{-\lambda_1 t} + e^{-\lambda_2 t} - e^{-(\lambda_1+\lambda_2)t} \tag{11.13}$$

The mean time between failures is

$$m_p = \int_0^\infty R_p \, dt = \int_0^\infty e^{-\lambda_1 t} \, dt + \int_0^\infty e^{-\lambda_2 t} \, dt - \int_0^\infty e^{-(\lambda_1+\lambda_2)t} \, dt$$

$$= \frac{1}{\lambda_1} + \frac{1}{\lambda_2} - \frac{1}{\lambda_1 + \lambda_2} \tag{11.14}$$

When the two parallel components in Figure 11.1 are equal so that

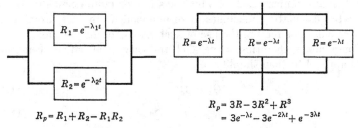

$$R_p = R_1 + R_2 - R_1 R_2$$

$$R_p = 3R - 3R^2 + R^3$$
$$= 3e^{-\lambda t} - 3e^{-2\lambda t} + e^{-3\lambda t}$$

Fig. 11.1. Reliability of two components in parallel.   **Fig. 11.2.** Three equal components in parallel.

$\lambda_1 = \lambda_2 = \lambda$, the unreliability of this parallel combination of two equal components is

$$Q_p = Q_1 Q_2 = Q^2 = (1 - e^{-\lambda t})^2 \tag{11.15}$$

The reliability is

$$R_p = 1 - Q_p = 1 - (1 - e^{-\lambda t})^2 = 2e^{-\lambda t} - e^{-2\lambda t} \tag{11.16}$$

The mean time between failures is

$$m_p = \frac{2}{\lambda} - \frac{1}{2\lambda} = \frac{1}{\lambda} + \frac{1}{2\lambda} = \frac{3}{2\lambda} \tag{11.17}$$

For three equal components in parallel, shown in the block diagram of Figure 11.2, we have

$$R_p = 1 - Q_p = 1 - Q^3 = 1 - (1 - e^{-\lambda t})^3 = 3e^{-\lambda t} - 3e^{-2\lambda t} + e^{-3\lambda t} \tag{11.18}$$

$$m_p = \frac{1}{\lambda} + \frac{1}{2\lambda} + \frac{1}{3\lambda} = \frac{11}{6\lambda} \tag{11.19}$$

* Redundant systems which are regularly inspected and maintained assume a constant average failure rate, as shown in Chapter 20.

When the three components in parallel are not equal,

$$R_p = 1 - Q_1 Q_2 Q_3 = 1 - (1 - e^{-\lambda_1 t})(1 - e^{-\lambda_2 t})(1 - e^{-\lambda_3 t}) \quad (11.20)$$

$$m_p = \frac{1}{\lambda_1} + \frac{1}{\lambda_2} + \frac{1}{\lambda_3} - \frac{1}{\lambda_1 + \lambda_2} - \frac{1}{\lambda_1 + \lambda_3} - \frac{1}{\lambda_2 + \lambda_3} + \frac{1}{\lambda_1 + \lambda_2 + \lambda_3}$$
$$(11.21)$$

Finally, for n equal components in parallel we obtain

$$R_p = 1 - Q_p = 1 - Q^n = 1 - (1 - e^{-\lambda t})^n \quad (11.22)$$

$$m_p = \frac{1}{\lambda} + \frac{1}{2\lambda} + \frac{1}{3\lambda} + \cdots + \frac{1}{n\lambda} \quad (11.23)$$

The instantaneous failure rates of parallel systems can be computed from R_p according to Equation (4.9). We shall make use of parallel system failure rates in Chapter 20.

The improvement in reliability achieved by operating components in parallel is shown in the following example, but it must be remembered that not all components are suitable for what we have defined as parallel operation, i.e., continuous operation of two parallel sets for the sole purpose of having one to carry on the operation alone should the other fail. Resistors and capacitors are particularly unsuitable for this kind of operation because if one fails out of two parallel units, this changes the circuit constants. When high reliability requirements make redundant arrangements of such units a necessity, these arrangements must then be of the stand-by type where only one unit operates at a time and the second unit, which is standing by idly, is switched into the circuit if the first unit fails. We shall deal with stand-by reliabilities in the next chapter.

Let us now assume that three equal units, each of which consists of a number of components in series, operate together as parallel units. The failure rate of each unit is the sum of the failure rates of its series components. Let this sum be $\lambda = 0.01$. The reliability of a single unit for a 10-hour operation is therefore

$$R = e^{-0.1} = 0.90484$$

or about 90 per cent. How much can we improve the reliability of this link, which may be a very vital link in a complex system, by operating three units in parallel?

The unreliability of a single unit for $t = 10$ hours is

$$Q = 1 - R = 1 - 0.90484 = 0.09516$$

The unreliability of the parallel group is

$$Q_p = Q^3 = (0.09516)^3 = 0.000862$$

The reliability of the parallel group of the three units for a 10-hour operation is then

$$R_p = 1 - Q_p = 1 - 0.000862 = 0.999138$$

or about 99.9 per cent. Thus, we have increased the reliability of this vital link in a complex system from 90 per cent to 99.9 per cent. The practical meaning of this improvement is that whereas a single unit would cause the complex system in which it operates to fail 10 times out of 100 operations, or 100 times out of 1000 operations, when we strengthen the link by putting three units in parallel and make sure that each 10-hour operation starts with all three units in good condition, we would expect this link to fail only about once in 1000 operations. This is because all three units would have to fail in a single 10-hour operation to cause the link—and therefore the system—to fail.

Let us also look at what would happen if we did not maintain this link of three parallel units after each operation, but let it operate until it fails, which means, until all three units fail in a succession of several operations. We now do not know whether one or two units have already failed, and we wait until the third unit fails when we replace all three of them. Would we still get only one failure of the link in 1000 ten-hour operations?

To get the answer we compute the mean time between failures of the three-unit parallel group:

$$m_p = \frac{1}{\lambda} + \frac{1}{2\lambda} + \frac{1}{3\lambda} = 100 + 50 + 33 = 183 \text{ hours}$$

On the average, we get a failure every 183 hours, or every 18.3 ten-hour operations. Therefore in 1000 operations we would expect the link to fail $1000/18.3$ times or about 54 times. This corresponds to a reliability of 946 successful operations and 54 unsuccessful operations out of a total of 1000 operations, and therefore to a link average reliability of 94.6 per cent for 10 hours. We say "average reliability" because obviously at the start of the operation when all three units are good, the reliability is 99.9 per cent for 10 hours; when two units are still good it is 99.1 per cent; but when just one unit is left it is only 90 per cent, so 0.946 is an average.

We obtain the same average result when we consider the nonmaintained link of the three parallel units to be a single nonexponential component which over long periods of time fails exponentially with a mean time between failures of $m = 183$ hours, because $e^{-10/183} = e^{-0.054} = 0.946$, for 10 hours on the average.

We see now that by neglecting to maintain the parallel group after each operation, we did not gain much reliability with a repeatedly operated or long-life system as compared to a single unit link which has a reliability of 0.905. But when we maintain the group after each operation so as to be certain that all three units are good, we achieve a reliability of 0.999138 for each 10-hour operation with a risk of only one failure of the group in 1000 operations. Thus, when reliability matters, we must maintain parallel reliabilities after each operation, which means we check every time whether one or perhaps two units might not have failed in an operation, and if there is a failed unit we must replace it before the next operation starts.

For comparative purposes we show in Figure 11.3 the reliability

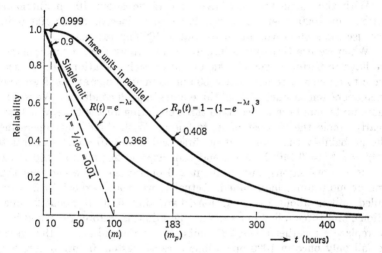

Fig. 11.3. The survival characteristic of three units in parallel. ($\lambda = 0.01$ per single unit.)

curves, or the survival characteristics, of the three parallel units and of a single unit. From this graph we see that a gain of several orders of magnitude is achieved in reliability for short stretches of operation when three units operate in parallel as compared with the exponential curve of a single unit. For instance, for 10 hours we have a reliability of 0.999 in the parallel case, but a reliability of only 0.90 for the single unit. This is a gain of three orders of magnitude. The graph also shows that the survival characteristic of a parallel combination of reliabilities is not exponential and therefore the failure rate is not constant. But the curve has a mean, which in this example is at 183 hours, whereas the mean of the single-unit exponential curve is at 100 hours. The value of R at $t = m$ is 0.37 for the exponential curve, but at $t = m_p$ it is 0.408 in the case of the three parallel units. This value increases with an increasing number of parallel units.

Thus parallel operation is capable of providing extremely high reliabilities for comparatively short operating periods. If we wanted a single unit to provide the same 0.999 reliability for 10 hours, such a unit would need a mean time between failures of 10,000 hours.

We have discussed the reliability aspects of the three operating schemes in the above example: using a single unit with $m = 100$ hours, using three of these units in parallel and maintaining them after each operation, and maintaining the three units only if all three fail. Now we shall briefly look into the maintenance aspect.

With the single unit we have to perform about 100 maintenance actions in 1000 operations. We have to replace or repair 100 units, because the system will fail in 100 out of 1000 operations.

When we use three units in parallel and apply perfect maintenance, we have to perform 1000 checks, one after each operation, and we shall have to replace or repair almost 300 units in 1000 operations if we want each operation to start with all three units in good condition. We replace more units here because we have three operating all the time, which will nearly treble the number of units failing. With three units operating, the probability that none out of three will fail in 10 hours amounts to $e^{-3\lambda t} = e^{-0.3} = 0.74082$. We would thus expect that no unit will fail in about 740 operations, but in about 260 operations we would find that one or more units have failed. Actually, we would expect to find one failed unit in about 234 operations, two failed units in about 25 operations, and three failed units in one operation; therefore, we would have to replace altogether about 287 units. But we would expect the system to fail only once in 1000 operations because system failure occurs only when all three units fail.

In the third scheme we operate three units in parallel but do not check them, and we wait until all three fail. This is a nonreplacement scheme. Here we expect 54 maintenance actions in 1000 operations with $3 \times 54 = 162$ units replaced or repaired. Thus, this system would fail on the average 54 times in 1000 operations.

We compare now the three operating schemes. Scheme 1 requires an average of 100 maintenance actions and 100 replacement units; the system fails on the average 100 times in 1000 operations. Scheme 2 requires 1000 checkings, and an average of 260 maintenance actions and 287 replacement units, but this system fails only once in 1000 operations. Scheme 3 requires an average of 54 maintenance actions and 162 replacement units; the system fails 54 times in 1000 operations. It is obvious, as far as reliability is concerned, that scheme 2 is far superior to the other two schemes. But it creates a real burden for maintenance. However, if a 0.999 reliability is required for a 10-hour mission and if no better units than those having $m = 100$ hours can be obtained, scheme 2 is the only solution.

Now, how much would it be worth to obtain a single unit with $m = 10,000$ hours? This unit would be expected to fail on the average once in 1000 operations (i.e., in 10,000 hours), and therefore only one maintenance action and one replacement unit would be needed. This single unit would fulfill the reliability requirement of 0.999 for 10 hours, and there would be almost no maintenance necessary. The cost of having such superior units built must be weighed against the cost of operating scheme 2. In that scheme we have the cost of $287 + 3 = 290$ units, plus the cost of 1000 checkings, plus the cost of 260 maintenance actions—in 234 one unit is replaced, in 25 two units are replaced, and in one three units are replaced. This total cost must be compared with the cost of $1 + 1$ units which have $m = 10,000$ hours, plus one maintenance action. It is not impossible that, for the same expense involved in the operation of scheme 2, such units could be built, and perhaps for even less. At least some worthwhile compromise could be achieved by a considerable increase in the mean time between failures. Equipment weight and volume must also be considered, as well as accessibility in scheme 2 for frequent checkings and maintenance.

In the discussion regarding the maintenance of scheme 2, we wanted to know how many units, which have a reliability of 0.905 each, will have to be replaced in 260 maintenance actions. This problem is equivalent to casting dice or drawing balls, where the binomial distribution is applicable. If we draw balls from a lot of 1000 balls of which 905 are white and 95 are black, and one trial consists of drawing three balls, we know that the probability of drawing all three white is $(^{905}/_{1000})^3$, or about 0.740, when we return each ball after a draw. The probability of drawing all three black is $(^{95}/_{1000})^3$, or about 0.001. Thus, in 1000 trials we would expect to draw three white balls 740 times and three black balls only once. As to the remaining 259 trials, there is a probability of $3(^{905}/_{1000})^2 \cdot (^{95}/_{1000}) = 0.234$ of drawing two white and one black in a single trial, or 234 such draws in 1000 trials. The "3" means that there are three possible arrangements of the succession of ball colors: white-white-black, white-black-white, or black-white-white. Similarly there is a probability of $3(^{905}/_{1000}) \cdot (^{95}/_{1000})^2 = 0.025$ of drawing one white and two black balls. This exhausts all the possibilities and therefore the sum of the probabilities must equal unity:

$$0.740 + 0.234 + 0.025 + 0.001 = 1$$

As the reader can see, this example is identical to operating three components and counting the survivals and failures in trials each of which lasts 10 hours. The probability of one component surviving 10 hours is $R = 0.905$ and its probability of not surviving 10 hours is $Q = 0.095$ so that $R + Q = 1$. This is the same as the probability of drawing a white ball in a single trial is 0.905, and the probability of not drawing a white ball (which equals the probability of drawing a black one) is 0.095 so that $0.905 + 0.095 = 1$.

If we make three draws in one trial or if we have three equal components in operation, we obtain the probabilities of all the possible outcomes of a trial or of an operation of a given length by expanding the binomial

$$(R + Q)^3 = R^3 + 3R^2Q + 3RQ^2 + Q^3 = 1 \qquad (11.24)$$

The first term represents the probability that all three components will survive the operation, the second term represents the probability that two will survive and one will fail, the third term is the probability that one will survive and two will fail, and the last term is the probability that all three will fail. We have defined this last term as the unreliability of the parallel system $Q_p = Q^3$ because as long as at least one component operates, the system does not fail. System reliability is therefore given by

$$R_p = R^3 + 3R^2Q + 3RQ^2 = 1 - Q_p \qquad (11.25)$$

If we were to specify that out of the three parallel components at least two must survive to prevent the system from failing, the term $3RQ^2$ would not be part of the system's reliability because we cannot tolerate the failure of two components in one operation. System unreliability would thus be

$$Q_p = 3RQ^2 + Q^3$$

and system reliability,

$$R_p = R^3 + 3R^2Q \qquad (11.26)$$

For four equal components operating in parallel, the binomial expansion reads

$$(R + Q)^4 = R^4 + 4R^3Q + 6R^2Q^2 + 4RQ^3 + Q^4 = 1 \quad (11.27)$$

To obtain the reliability of this parallel system we discard those terms in the expanded binomial which represent system failure. If at least one component must survive to prevent the system from failing, only the last term Q^4 represents system failure, and

$$R_p = R^4 + 4R^3Q + 6R^2Q^2 + 4RQ^3$$

If at least three components out of four must survive to complete a mission successfully, $R_p = R^4 + 4R^3Q$, because the failure of two components ($6R^2Q^2$) and the failure of three components ($4RQ^3$) represent system failures. Therefore, the system unreliability is then

$$Q_p = 6R^2Q^2 + 4RQ^3 + Q^4$$

Situations such as this exist in four-engine aircrafts where the mission can be completed with one engine having failed, but when two engines fail an emergency landing must be made. Also, missile propulsion systems with a number of clustered engines such as the Saturn missile sometimes allow the failure of one or more of the engines.

If there are n equal components or units in a parallel system, the binomial expansion is

$$(R + Q)^n = R^n + nR^{n-1}Q + \frac{n(n-1)}{2!} R^{n-2}Q^2 + \frac{n(n-1)(n-2)}{3!}$$

$$R^{n-3}Q^3 + \frac{n(n-1)(n-2)(n-3)}{4!} R^{n-4}Q^4 + \cdots + Q^n = 1 \quad (11.28)$$

where R and Q are the reliability and the unreliability of one component or unit.

To apply the binomial distribution, R, and therefore Q, must be known for a definite operating time. We can then calculate the probabilities of various possible combinations of events in that particular time period, which means that the parallel components or units have to operate simultaneously in a time period of given length, and that they must be equal or must have the same reliability.

If the components are not equal but have different reliabilities, the calculation becomes more complicated. Assume three components with the reliabilities R_1, R_2, and R_3 operating simultaneously and in parallel. We want to know the reliability of this system when at least one of them must survive, and when two of them must survive to make the operation a success. We then proceed by developing, instead of $(R + Q)^3$, the term

$$(R_1 + Q_1)(R_2 + Q_2)(R_3 + Q_3) = R_1R_2R_3 + (R_1R_2Q_3 + R_1R_3Q_2 + R_2R_3Q_1) + (R_1Q_2Q_3 + R_2Q_1Q_3 + R_3Q_1Q_2) + Q_1Q_2Q_3 = 1 \quad (11.29)$$

and to obtain R_p of the system, according to the specified conditions we discard the last term only, i.e., $Q_1Q_2Q_3$, or the last term and the terms in brackets before it which contain the product of two unreliabilities. Similarly, for four unequal components in parallel we would expand

$$(R_1 + Q_1)(R_2 + Q_2)(R_3 + Q_3)(R_4 + Q_4) = 1$$

and discard the terms which represent system failure. We shall see later that there are situations where such complicated calculations must be performed.

Cases of straightforward binomial expansion occur in aircraft and missile propulsion systems, electrical generating systems, etc., where equal units operate in parallel. As an example, we consider the reliability of an aircraft propulsion system which consists of three engines.

Each engine consists of a multitude of components in series with failure rates λ_i so that the failure rate of an engine is $\lambda = \Sigma \lambda_i = 0.0005$. It is necessary to calculate the reliability of the propulsion system for a mission length of 10 hours when at least two engines must survive to make the mission a success. Figure 11.4 shows the reliability block diagram of the system. The reliability of one engine for 10 hours is

$$R = e^{-0.0005 \cdot 10} = e^{-0.005} = 0.995012$$

The binomial expansion for at least two units surviving out of three operating in parallel results in Equation (11.26), and therefore the reliability of this propulsion system is

$$R_p = R^3 + 3R^2Q = R^3 + 3R^2(1 - R) = 3R^2 - 2R^3$$
$$= 3(0.995012)^2 - 2(0.995012)^3 = 0.999926$$

or better than 99.99 per cent—a very high reliability. However, if the mission has to last 100 hours, the reliability of one engine becomes about $R = 0.95$ and the system reliability for 100 hours reduces to about $R_p = 0.99$ or 99 per cent; if the mission time were 1000 hours, system reliability would be only about 0.6 or 60 per cent. Of course powered flights of such duration would be possible only with nuclear power plants.

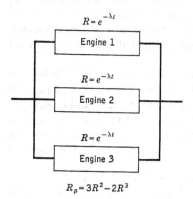

Fig. 11.4. The reliability of three aircraft engines in parallel.

Finally, let us consider the effects of partial redundancy on system reliability. Under partial redundancy we understand the paralleling of certain individual components or groups of components in a system.

It follows from Equation (11.22) that the reliability of n equal components in parallel is

$$R_p = 1 - (1 - r)^n \tag{11.30}$$

where r is the reliability of one component. If we parallel groups of components—for instance, circuits—each having m components in series, so that the reliability of one circuit is

$$R = r_1 r_2 r_3 \cdots r_m = \prod_{i=1}^{m} r_i$$

we obtain for n equal circuits in parallel,

$$R_p = 1 - (1 - R)^n = 1 - (1 - \prod_{i=1}^{m} r_i)^n \tag{11.31}$$

If $n = 2$, i.e., if we parallel two equal circuits with m series components each, we obtain

$$R_p = 1 - (1 - \prod_{i=1}^{m} r_i)^2 \qquad (11.32)$$

The paralleling of two equal groups of components (such as circuits, modules, units, or any other building blocks) is often done in reliability design work. However, in exceptional cases the paralleling of two such units still may not provide the required reliability, and it may become necessary to parallel three units. Before making this decision it is worthwhile to look into the possibility of paralleling some components in each of the units, which would increase the reliability of each unit. It may then become possible to obtain the required system reliability by paralleling just two units. This procedure can save equipment weight and volume, and may also save on system maintenance if accessibility is duly considered during the design stage.

In a unit which at first consists of m components in series, when we decide to duplicate individually b components so that when one of the duplicated components fails its parallel operating partner component will continue to perform the required function, and we leave the remaining $a = m - b$ components as series elements in the unit, the reliability of the unit is given by

$$R = (\prod_{i=1}^{a} r_i) \cdot \prod_{j=1}^{b} [1 - (1 - r_j)^2] \qquad (11.33)$$

In this equation r_i stands for the reliability of the ith unparalleled component, and r_j for the reliability of the jth individually paralleled component. The reliability of a system of two of these units is then

$$R_p = 1 - (1 - R)^2 \qquad (11.34)$$

where R is given by Equation (11.33). This method of partial redundancy effects considerable increases in the reliability of parallel arrangements at minimum cost and effort.

We can rewrite Equations (11.32) and (11.34) for exponential components as follows:

$$R_p = 1 - [1 - \exp(- \sum_{i=1}^{m} \lambda_i t)]^2 \qquad (11.35)$$

which is the reliability of two equal series units in parallel, and

$$R_p = 1 - \{1 - \exp(- \sum_{i=1}^{a} \lambda_i t) \cdot \prod_{j=1}^{b} [1 - (1 - e^{-\lambda_j t})^2]\}^2 \qquad (11.36)$$

which is the reliability of two units in parallel, each of which contains partial redundancy in the form of b of its components being individually duplicated.

Obviously, partial redundancy increases the number of components in a unit. Originally, when the unit was a purely series system, it contained $m = a + b$ components; after the duplication of b components it contained $a + 2b$ components. However, its reliability increased from that given by Equation (11.35) to that given by Equation (11.36).

Sometimes it is more practical to duplicate an entire string of components in a unit than to duplicate each component individually. However, the increase in reliability is then less pronounced than in the case of individual component duplication, because should any one of the components in a string fail, the entire string is put out of action even though it still contains other good components—these good components in the string are then useless. And if one component fails also in the parallel string, the whole system fails, whereas when we duplicated each component individually, one component of each double of components could fail and the system would still operate.

The reliability of a unit in which a nonduplicated components are in series and b components are duplicated as one string and not individually, so that the unit contains $a + 2b$ components, is in the exponential case

$$R = \exp\left(-\sum_{i=1}^{a} \lambda_i t\right) \cdot \{1 - [1 - \exp\left(-\sum_{j=1}^{b} \lambda_j t\right)]^2\} \qquad (11.37)$$

and in general,

$$R = \left(\prod_{i=1}^{a} r_i\right) \cdot [1 - (1 - \prod_{j=1}^{b} r_j)^2] \qquad (11.38)$$

which is less than the reliability of the same unit with b individually duplicated components as given by Equation (11.33).

When we parallel two of these units which contain duplicated strings, we obtain

$$R_p = 1 - \{1 - \exp\left(-\sum_{i=1}^{a} \lambda_i t\right) \cdot [1 - \{1 - \exp\left(-\sum_{j=1}^{b} \lambda_j t\right)\}^2]\}^2 \qquad (11.39)$$

or, in general,

$$R_p = 1 - \{1 - \left(\prod_{i=1}^{a} r_i\right) \cdot [1 - (1 - \prod_{j=1}^{b} r_j)^2]\}^2 \qquad (11.40)$$

Although this is a lesser reliability than that given by Equation (11.36), an improvement is nevertheless achieved over the reliability given by Equation (11.35). Which approach is chosen depends from case to case on the system reliability requirements, and on the design possibilities of duplicating components individually or groups of components, such as subcircuits, or circuits. As every designer knows, it is often impossible to have individual components operate in parallel reliability (for instance, resistors or capacitors). Also, the parallel redundant operation of circuits often leads to difficulties. In such cases the approach of stand-by redundancy must be chosen; this will be discussed in the next chapter.

As seen above, every introduction of redundancy increases the number of components in a system. When the system is a recoverable one which is subject to regular and repair maintenance, more components usually mean more maintenance effort. This may not apply to one-shot systems where reliability is the main concern.

The increase of maintenance follows from the following consideration: If a unit consists of m exponential components, its probability P_M that it will not require maintenance in time t equals the probability that none of its m components will fail in time t:

$$P_M = \exp\left(-\sum_{i=1}^{m} \lambda_i t\right) \tag{11.41}$$

This is also the reliability of a series unit or system with m components.

If the unit has a components in series and b duplicated components, so that it contains $a + 2b$ components (as compared to the $m = a + b$ components of a series unit), its probability that none of the $a + 2b = m + b$ components will fail in time t is

$$P_M = \exp\left[-\left(\sum_{i=1}^{a} \lambda_i + \sum_{j=1}^{b} 2\lambda_j\right)t\right] \tag{11.42}$$

This is a lower probability of maintenance-free operation than that of the series unit, but the reliability of the unit is now much higher than that of the series unit and is given by Equations (11.37) or (11.33), according to whether the b components are duplicated individually or as a string. Obviously, when the unit contains redundancy, its probability that none of its components will fail in time t is no longer equal to the unit's reliability.

It follows from Equations (11.41) and (11.42) that the probability P_M that no component will fail in time t is obtained from the simple exponential formula by adding up the failure rates of all system components, whether redundant or not.

If the components do not behave exponentially, P_M becomes the product of the reliabilities R_i of all system components for the period t for which P_M is to be evaluated:

$$P_M = \Pi R_i \tag{11.43}$$

In this equation R_i, the reliability of the ith component, will in general be a function of both the operating period t and the component age T, therefore $R_i(t, T)$, as given by Equation (8.23).

Chapter 12

RELIABILITY OF STAND-BY
SYSTEMS

OFTEN IT IS NOT FEASIBLE or practical to operate components or units in parallel and so-called "stand-by" arrangements must be applied; that is, when a component or unit is operating, one or more components or units are standing by to take over the operation when the first fails. Whereas in a parallel operation all units operate simultaneously, in the stand-by case the supporting units are standing by idly and begin to operate only when the preceding unit fails.

Stand-by arrangements normally require failure-sensing and switchover devices to put the next unit into operation. Let us first assume that the sensing and switchover devices are 100 per cent reliable and that the operating component and the stand-by components have the same constant failure rate λ.

We can regard such a group of stand-by components as being a single unit or system which is allowed to fail a number of times before it definitely stops performing its function. If n components are standing by to support one operating component, we have $n + 1$ components in the system, and n failures can occur without causing the system to fail. Only the $(n + 1)$th failure would cause system failure.

Making use of Equations (9.13) through (9.17), we know that the following identity is true:

$$e^{-\lambda t}\left(1 + \lambda t + \frac{(\lambda t)^2}{2!} + \frac{(\lambda t)^3}{3!} + \cdots\right) = 1 \qquad (12.1)$$

In this expression the term $e^{-\lambda t} \cdot 1$ represents the probability that no failure will occur, the term $e^{-\lambda t}(\lambda t)$ represents that probability that exactly one failure will occur, $e^{-\lambda t}(\lambda t)^2/2!$ represents the probability that exactly two failures will occur, etc. Therefore, the probability that one failure or no failure will occur equals $e^{-\lambda t} + e^{-\lambda t}(\lambda t)$, the probability that two or one or no failure will occur or the probability that not more than two failures will occur equals $e^{-\lambda t} + e^{-\lambda t}(\lambda t) + e^{-\lambda t}(\lambda t)^2/2!$, etc. As the reader may have already guessed, these probabilities represent the reliability of various stand-by arrangements of equal components, i.e., of components which have the same failure rate.

If we denote by R_b and Q_b the reliability and the unreliability of a whole stand-by system, and because $R_b + Q_b = 1$, we can write

$$R_b + Q_b = e^{-\lambda t}\left(1 + \lambda t + \frac{(\lambda t)^2}{2!} + \frac{(\lambda t)^3}{3!} + \cdots\right)$$

$$= e^{-\lambda t} + (\lambda t)e^{-\lambda t} + \frac{\lambda^2 t^2}{2!}e^{-\lambda t} + \cdots = 1 \qquad (12.2)$$

If in this expanded term we allow one failure, then

$$R_b = e^{-\lambda t} + e^{-\lambda t}(\lambda t) = e^{-\lambda t}(1 + \lambda t) \qquad (12.3)$$

which is the reliability of a stand-by system composed of one operating component or series of components, with a component or series failure rate λ, and of another component or series with the same failure rate and standing by idly to take over if the first fails. This reliability is the probability that one or no failure will occur. The unreliability Q_b of the whole stand-by system is then the probability that at least two failures occur:

$$Q_b = 1 - R_b = 1 - e^{-\lambda t}(1 + \lambda t) = e^{-\lambda t}\left[\frac{(\lambda t)^2}{2!} + \frac{(\lambda t)^3}{3!} + \frac{(\lambda t)^4}{4!} + \cdots\right] \quad (12.4)$$

The mean time between failures of the system is obtained by the integration of R_b, and therefore for a two-component system we obtain

$$m_b = \int_0^\infty R_b\,dt = \int_0^\infty e^{-\lambda t}\,dt + \int_0^\infty e^{-\lambda t}(\lambda t)\,dt = \frac{1}{\lambda} + \frac{\lambda}{\lambda^2} = \frac{2}{\lambda} \quad (12.5)$$

But, as in the case of parallel operating components, the failure rate of a stand-by system is not constant and the reliability function, or survival characteristic $R_b(t)$, is not exponential but resembles the graph in Figure 11.3.

For a stand-by system of three units which have the same failure rate and where one unit is operating and the other two are standing by to take over the operation in succession, we have

$$R_b = e^{-\lambda t}\left(1 + \lambda t + \frac{\lambda^2 t^2}{2!}\right) \qquad (12.6)$$

$$m_b = \frac{1}{\lambda} + \frac{1}{\lambda} + \frac{1}{\lambda} = \frac{3}{\lambda} \qquad (12.7)$$

And, in general, when n equal components or units are standing by to support one which operates,

$$R_b = e^{-\lambda t}\left(1 + \lambda t + \frac{\lambda^2 t^2}{2!} + \cdots + \frac{\lambda^n t^n}{n!}\right) \qquad (12.8)$$

$$m_b = \frac{n+1}{\lambda} \qquad (12.9)$$

As an example we calculate the reliability for $t = 10$ hours of a stand-by system consisting of two equal units, each with $\lambda = 0.01$:

$$R_b = e^{-\lambda t} + e^{-\lambda t}(\lambda t) = e^{-0.1} + e^{-0.1}(0.1)$$

$$= 0.90484 + 0.90484(0.1) = 0.995324$$

whereas the reliability of a single unit is 0.90484 and the reliability of a parallel arrangement of two units is 0.990945. Thus, we see that stand-by arrangements are only slightly more reliable than parallel operating units, although they have a considerably longer mean time between failures. However, these advantages are easily lost when the reliability of the sensing-switching device R_{ss} is less than 100 per cent, which is normally the case. Taking this into consideration, and when the circuits are arranged so that the reliability of the operating unit is not affected by the unreliability of the sensing-switching device, we obtain for a system in which one stand-by unit is backing up one operating unit:[*]

$$R_b = e^{-\lambda t} + R_{ss}e^{-\lambda t}\,\lambda t \qquad (12.10)$$

It is the exception rather than the rule that the failure rates of the stand-by units are equal to those of the operating unit. For instance, a hydraulic actuator will be backed up by an electrical actuator, and there may be even a third stand-by unit, pneumatic or mechanical. In such cases the failure rates of the stand-by units will not be equal and the formulas which we derived above will no longer apply.

Another approach is needed—it consists of first deriving the density function $f(t)$ of a given combination of components in stand-by, and obtaining the cumulative reliability of the system by the integration of the density function as

$$R_b = \int_t^\infty f(t)\,dt$$

This procedure is always permissible and correct, whether or not the components are equal or fail exponentially.

We have seen in Chapter 4 that the exponential density function is $f(t) = \lambda e^{-\lambda t}$. This applies to a single component as well as a series system

---

[*] R_{ss}, here, is assumed to be independent of time and refers to a single switch-over operation when the first unit fails.

with a constant failure rate λ. We shall now form the density function of a system consisting of two components or units, with the failure rates λ_1 and λ_2, the one operating and the other standing by. The system is such that if the operating component fails at a time t_1, the stand-by component begins to operate immediately. The time to failure of the stand-by component is $t_2 = t - t_1$ when the operating time t_2 of this component is counted from the moment t_1 when the first component has failed. Obviously, both times t_1 and t are variables. Therefore, the density function of the first component is

$$f_1(t_1) = \lambda_1 e^{-\lambda_1 \cdot t_1}$$

and the density function of the second component is

$$f_2(t_2) = \lambda_2 e^{-\lambda_2(t-t_1)}$$

From the density functions of the components we derive their probabilities of failure in infinitely small intervals dt, and knowing that system failure occurs only if both components fail, we obtain the system probability of failure by multiplication.

Thus, we have for the first component $\lambda_1 e^{-\lambda_1 t_1}\, dt_1$ and for the second component $\lambda_2 e^{-\lambda_2(t-t_1)}\, dt$, and the probability of system failure in a small interval from t to $t + dt$ is then the product

$$\lambda_1 \lambda_2 e^{-\lambda_1 t_1} e^{-\lambda_2(t-t_1)}$$

We have two variable times, t_1 and t, in this product, and if we integrate with respect to t_1 we obtain the joint density function $f(t)$ of the stand-by system expressed in terms of a single variable t, because $t_2 = t - t_1$:

$$f(t) = \int_{t_1=0}^{t} f_1(t_1) f_2(t_2)\, dt_1 \tag{12.11}$$

for two components of which the first is operating and the second is standing by idly. If f_1 and f_2 are not exponential, then $f(t)$ will be dependent on the age of the components, and the reliability of such stand-by systems for a given operating interval is then calculated from the conditional probability of failure as shown in Chapter 6, Equations (6.8) and (6.10).

In the exponential case, however, the calculation is simplified by the independence from the component age. The joint density function of two unequal exponential components is then

$$f(t) = \lambda_1 \lambda_2 \int_{t_1=0}^{t} e^{-\lambda_1 t_1} e^{-\lambda_2(t-t_1)}\, dt_1$$

$$= \frac{\lambda_1 \lambda_2}{\lambda_2 - \lambda_1} (e^{-\lambda_1 t} - e^{-\lambda_2 t})$$

$$= \lambda_1 \lambda_2 \left(\frac{e^{-\lambda_1 t}}{\lambda_2 - \lambda_1} + \frac{e^{-\lambda_2 t}}{\lambda_1 - \lambda_2} \right) \tag{12.12}$$

The reliability for an operating time t is then obtained by integration of the density function as follows:

$$R_b(t) = \int_t^\infty f(t)\, dt = \frac{\lambda_2}{\lambda_2 - \lambda_1} e^{-\lambda_1 t} + \frac{\lambda_1}{\lambda_1 - \lambda_2} e^{-\lambda_2 t}$$

$$= e^{-\lambda_1 t} + \frac{\lambda_1}{\lambda_2 - \lambda_1} (e^{-\lambda_1 t} - e^{-\lambda_2 t}) \tag{12.13}$$

And the mean time between failures becomes

$$m_b = \int_0^\infty R_b(t)\, dt = \frac{1}{\lambda_1} + \frac{1}{\lambda_2} = m_1 + m_2 \tag{12.14}$$

If the reliability of the sensing-switching device is not 100 per cent, Equation (12.13) becomes

$$R_b(t) = e^{-\lambda_1 t} + R_{ss} \frac{\lambda_1}{\lambda_2 - \lambda_1} (e^{-\lambda_1 t} - e^{-\lambda_2 t}) \tag{12.15}$$

Equation (12.11) for the joint density function can be extended to three components of which one is operating and two are standing by to take over the operation in succession:

$$f(t) = \int_{t_2=0}^t \int_{t_1=0}^{t_2} f_1(t_1) f_2(t_2) f_3(t_3)\, dt_1\, dt_2 \tag{12.16}$$

where t_1 stands for the time to failure of the operating component, t_2 stands for the time to failure of the first stand-by component, and the second stand-by component has a time to failure $t > t_2$.

In the case of three exponential components with failure rates λ_1, λ_2, and λ_3, the joint density function is

$$f(t) = \lambda_1 \lambda_2 \lambda_3 \int_{t_2=0}^t \int_{t_1=0}^{t_2} e^{-\lambda_1 t_1} e^{-\lambda_2 (t_2 - t_1)} e^{-\lambda_3 (t - t_2)}\, dt_1\, dt_2 \tag{12.17}$$

By the integration of this density function we obtain the reliability of three components in stand-by as $\int_t^\infty f(t)\, dt$:

$$R_b(t) = \frac{\lambda_2 \lambda_3 e^{-\lambda_1 t}}{(\lambda_2 - \lambda_1)(\lambda_3 - \lambda_1)} + \frac{\lambda_1 \lambda_3 e^{-\lambda_2 t}}{(\lambda_1 - \lambda_2)(\lambda_3 - \lambda_2)} + \frac{\lambda_1 \lambda_2 e^{-\lambda_3 t}}{(\lambda_1 - \lambda_3)(\lambda_2 - \lambda_3)} \tag{12.18}$$

In a similar way we can form the joint density function for four and for n components, of which one is operating and $n - 1$ components are in stand-by each to the previous component, and derive from this the reliability by integration.

Thus, the reliability of a stand-by group of four components, with one operating and the others standing by, is

$$R_b(t) = \frac{\lambda_2\lambda_3\lambda_4 e^{-\lambda_1 t}}{(\lambda_2 - \lambda_1)(\lambda_3 - \lambda_1)(\lambda_4 - \lambda_1)} + \frac{\lambda_1\lambda_3\lambda_4 e^{-\lambda_2 t}}{(\lambda_1 - \lambda_2)(\lambda_3 - \lambda_2)(\lambda_4 - \lambda_2)}$$

$$+ \frac{\lambda_1\lambda_2\lambda_4 e^{-\lambda_3 t}}{(\lambda_1 - \lambda_3)(\lambda_2 - \lambda_3)(\lambda_4 - \lambda_3)} + \frac{\lambda_1\lambda_2\lambda_3 e^{-\lambda_4 t}}{(\lambda_1 - \lambda_4)(\lambda_2 - \lambda_4)(\lambda_3 - \lambda_4)} \quad (12.19)$$

The pattern of equations is now obvious, and therefore the reliability of a group of n components of which again one is operating and $n - 1$ are standing by becomes*

$$R_b(t) = \frac{\lambda_2\lambda_3\lambda_4 \cdots \lambda_n e^{-\lambda_1 t}}{(\lambda_2 - \lambda_1)(\lambda_3 - \lambda_1) \cdots (\lambda_n - \lambda_1)} + \frac{\lambda_1\lambda_3\lambda_4 \cdots \lambda_n e^{-\lambda_2 t}}{(\lambda_1 - \lambda_2)(\lambda_3 - \lambda_2) \cdots (\lambda_n - \lambda_2)}$$

$$+ \cdots + \frac{\lambda_1\lambda_2 \cdots \lambda_{i-1}\lambda_{i+1} \cdots \lambda_n e^{-\lambda_i t}}{(\lambda_1 - \lambda_i) \cdots (\lambda_{i-1} - \lambda_i)(\lambda_{i+1} - \lambda_i) \cdots (\lambda_n - \lambda_i)}$$

$$+ \cdots + \frac{\lambda_1\lambda_2\lambda_3 \cdots \lambda_{n-1} e^{-\lambda_n t}}{(\lambda_1 - \lambda_n)(\lambda_2 - \lambda_n) \cdots (\lambda_{n-1} - \lambda_n)} \quad (12.20)$$

The mean time between failures for n components in stand-by is then

$$m_b = \frac{1}{\lambda_1} + \frac{1}{\lambda_2} + \frac{1}{\lambda_3} + \cdots + \frac{1}{\lambda_n} \quad (12.21)$$

Equations (12.17) through (12.21) apply to the case of perfect switching and to strictly exponential components which cannot fail for any reason other than chance. In practice this would imply the replacement of the operating components when they are approaching a wearout condition, even if they have not yet failed. However, with several components in stand-by, one could allow the operating component to wear out if it does not fail of chance. In such case the component density function becomes one of combined chance and wearout failure probabilities, as shown in Chapters 6, 7, and 8.

In all stand-by problems discussed above, the assumption was made that the stand-by components have zero failure rates while waiting idly until the preceding component fails. However, this is not necessarily the case because idling components in an operating system can fail if exposed to high temperature or vibration levels. If this possibility exists, the idling components will also have some failure rate, which will probably be lower than their operating failure rate. In developing the equations for these situations, it is also necessary to include the probability of failure of the idling components during the period when they are not operating. The final formulas for two such components in stand-by are given on the next page.

* For the development of the equations for unequal components in stand-by, the author is indebted to Mrs. Julie Foster, mathematician with the Boeing Airplane Company in Renton, Washington.

Consider two unequal exponential components in stand-by, so that the operating component has a failure rate of λ_1, the stand-by component has a failure rate of λ_2 after it begins to operate, but also has a failure rate of λ_3 while standing by idly. The reliability of such a group is given by

$$R_b(t) = e^{-\lambda_1 t} + \frac{\lambda_1}{(\lambda_1 + \lambda_3) - \lambda_2} [e^{-\lambda_2 t} - e^{-(\lambda_1 + \lambda_3)t}] \qquad (12.22)$$

and if the reliability of the sensing-switching device is not unity,

$$R_b(t) = e^{-\lambda_1 t} + R_{ss} \frac{\lambda_1}{(\lambda_1 + \lambda_3) - \lambda_2} [e^{-\lambda_2 t} - e^{-(\lambda_1 + \lambda_3)t}] \qquad (12.23)$$

Let us now consider a d-c power supply—for instance, a generator—with a failure rate of $\lambda_1 = 0.0002$ and a stand-by battery with a failure rate of $\lambda_2 = 0.001$ when in operation. Assume that the switching circuit has a known reliability of $R_{ss} = 0.99$ for one switching operation. The reliability block diagram is shown in Figure 12.1. What is the reliability of this system for a mission time of t hours and for $t = 10$ hours?

$$R_b = e^{-0.0002t} + 0.99 \frac{0.0002}{0.001 - 0.0002} (e^{-0.0002t} - e^{-0.001t})$$

$$= e^{-0.0002t} + 0.2475(e^{-0.0002t} - e^{-0.001t})$$

$$= 0.998 + 0.2475(0.008) = 0.99998$$

for $t = 10$ hours. With this reliability, which amounts to a chance of failure of only two in 100,000 or one in 50,000, this system would probably be acceptable, when it is made certain that each mission begins with all components in good condition, and assuming that the failure rate of the battery is zero while it is standing by idling. What would be the effect on the reliability of this system if we were to decide not to check all components after each mission and to allow the system to operate without maintenance until it fails of chance? To get the answer we compute the system's mean time between failures by integration of R_b:

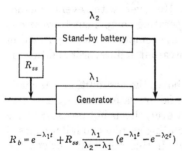

$$R_b = e^{-\lambda_1 t} + R_{ss} \frac{\lambda_1}{\lambda_2 - \lambda_1} (e^{-\lambda_1 t} - e^{-\lambda_2 t})$$

Fig. 12.1. Stand-by reliability.

$$m_b = \int_0^\infty R_b \, dt = \int_0^\infty \left[e^{-\lambda_1 t} + e^{-\lambda t} \frac{\lambda_1}{\lambda_2 - \lambda_1} (e^{-\lambda_1 t} - e^{-\lambda_2 t}) \right] dt$$

$$= \frac{1}{\lambda_1} + \frac{\lambda_1}{\lambda_2 - \lambda_1} \left(\frac{1}{\lambda_1 + \lambda} - \frac{1}{\lambda_2 + \lambda} \right)$$

where λ is the average failure rate of the switching circuit. This we obtain from the knowledge that the generator, which has a failure rate of 0.0002, can be expected to fail on the average once every 5000 hours, and therefore, the switching circuit would be required to operate on the average only once every 5000 hours. Its equivalent per-hour failure rate is thus

$$\lambda = - \ln \frac{0.99}{5000} = 0.000002$$

Substituting this value in the equation for m_b, we obtain for the stand-by system a mean time between failures of approximately 5988 hours which, based on a 10-hour mission, amounts to a chance of failure of one in 598—quite a difference compared with one in 50,000 when the system is maintained after each mission.

So far we have discussed cases where one or more units are in stand-by to support one operating unit. Another mode of stand-by occurs when a number of equal units are operating and one or more such units are standing by to take over the function of any failed unit. A typical case is that of a large number of equal electron tubes or semiconductors operating in series in an electronic circuit and n spares are available to the operating personnel as stand-by. If a tube fails it can be quickly replaced, and such tube failure is therefore, by agreement, not considered to be a system failure. Complete breakdown of operation occurs only when no more spares are available to replace a failed tube.

With n spares we thus allow n failures to occur in the system without causing complete system failure—only the $(n + 1)$th failure will constitute system failure. Therefore, the expected time for the occurrence of $n + 1$ failures is the system's mean time between failures. By the term "system" we understand here the series of the equal components or units. If the system consists of N such components in series, with a failure rate of λ each, the system failure rate is $N\lambda$, and the system mean time between failures is $1/N\lambda$ if no stand-by components are available. However, when we carry n spares as stand-by and quick replacement is possible, n failures can occur without system breakdown. The system mean time between failures therefore becomes $(n + 1)/N\lambda$, and system reliability is

$$R_s(t) = e^{-N\lambda t}\left[1 + N\lambda t + \frac{(N\lambda t)^2}{2!} + \cdots + \frac{(N\lambda t)^n}{n!} \right] \quad (12.24)$$

For instance, assume an airborne radar with 30 equal tubes, each having a failure rate of $\lambda = 0.001$ and assume that all other components are 100 per cent reliable. The system reliability for a 10-hour flight will be

$$R = \exp\left(-0.001 \cdot 30 \cdot 10\right) = \exp\left(-0.3\right) = 0.74082$$

But if three spare tubes are carried in stand-by by the operating personnel and the equipment is so built that replacements can be made for all practical purposes almost immediately—which implies built-in failure indicating and locating circuitry—the reliability of the system improves to

$$R = e^{-0.3}[1 + 0.3 + 0.045 + 0.0045] = 0.99973$$

Thus, a considerable increase in system reliability can be achieved. With a given reliability requirement it is then possible to specify the minimum number of spares which have to be carried on each mission. Such calculations may be of importance for long military mission flights, and manned orbiting and space vehicles where the weight of any spare equipment must be minimized but adequate reliability is required.

Chapter 13

BAYES' THEOREM IN RELIABILITY

NOT ALL RELIABILITY PROBLEMS can be reduced to the series, parallel, and stand-by models discussed in Chapters 10, 11, and 12. There exist combinations of components which are neither series, nor parallel, nor stand-by in the sense defined earlier.

Consider the schematic reliability block diagram in Figure 13.1. Two

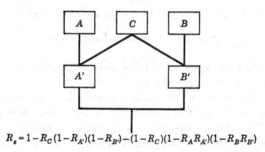

$$R_s = 1 - R_C(1 - R_{A'})(1 - R_{B'}) - (1 - R_C)(1 - R_A R_{A'})(1 - R_B R_{B'})$$

Fig. 13.1. Example of Bayes' theorem in reliability.

equal paths A-A' and B-B' operate in parallel so that if at least one of them is good the output is assured. However, because the units A and B are not reliable enough, a third equal unit C is inserted into the circuit so that it can supply either A' or B' with the necessary signal. Therefore, the following operations are possible: A-A', C-A', C-B', and B-B'. What is the reliability of this component arrangement?

If the circuit were connected as shown in Figure 13.2, it would be simple to solve the problem by evaluating first the reliability of the upper row consisting of A, C, and B in parallel, and multiplying this by the parallel reliability of the lower row A', B'.

Thus, Figure 13.2 is a typical case of two parallel reliabilities connected in series. The series connection is borne out by the fact that each of the components A, C, and B can by itself operate either with A' or with B'. However, this is not the case in the problem of Figure 13.1.

To solve the problem we use a simplified form of Bayes' probability theorem which says: If A is an event which depends on one of two mutually exclusive events B_i and B_j of which one must necessarily occur, then the probability of the occurrence of A is given by

$$P(A) = P(A, \text{ given } B_i) \cdot P(B_i) + P(A, \text{ given } B_j) \cdot P(B_j) \quad (13.1)$$

Translating this theorem into the language of reliability, and denoting by A the event of a system's failure and by B_i and B_j the survival and

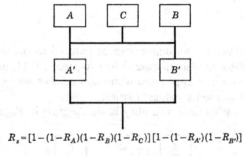

$$R_s = [1 - (1 - R_A)(1 - R_B)(1 - R_C)] [1 - (1 - R_{A'})(1 - R_{B'})]$$

Fig. 13.2. A parallel reliability case.

the failure of a component or unit on whose operation the system reliability depends, we can write the following rule:*

The probability of system failure equals the probability of system failure given that a specified component in the system is good, times the probability that the component is good, plus the probability of system failure given that the said component is bad, times the probability that the component is bad, or P(system failure) is equal to

$$P(\text{system failure if component } X \text{ is good}) \times P(X \text{ is good})$$
$$+ P(\text{system failure if } X \text{ is bad}) \times P(X \text{ is bad}) \quad (13.2)$$

In this equation X is a component or unit upon whose state (good or bad) the system reliability depends. Denoting by Q_s the probability of system failure, by R_x the probability that component X is good, and by

---

* The author is indebted to Dr. Judah Rosenblatt, Professor of Mathematics at Purdue University, Lafayette, Ind., for an acquaintance with the application of Equation (13.1) in reliability calculations. At that time (summer, 1957) Dr. Rosenblatt was associated with the Boeing Airplane Co., Renton, Wash.

Q_x the probability that it is bad, we obtain Equation (13.2) in the following conventional notation form:

$$Q_s = Q_s(\text{if } X \text{ is good}) \cdot R_x + Q_s(\text{if } X \text{ is bad}) \cdot Q_x \qquad (13.3)$$

System reliability is then

$$R_s = 1 - Q_s \qquad (13.4)$$

Equation (13.3) supplies us with one of the most powerful tools for the calculation of complex system reliabilities. We shall show that this equation is of a very general character, but let us first solve the problem given in Figure 13.1.

We pick component C for X in Equation (13.3) and write

$$Q_s = Q_s(\text{if } C \text{ is good}) \cdot R_c + Q_s(\text{if } C \text{ is bad}) \cdot Q_c$$

If C is good the system can fail only if both A' and B' fail, and because A' and B' are in parallel, the system's unreliability if C is good amounts to

$$Q_s(\text{if } C \text{ is good}) = (1 - R_{A'})(1 - R_{B'})$$

If C is bad the system can fail only if both parallel paths A-A' and B-B' fail, and the system's unreliability if C is bad amounts to

$$Q_s(\text{if } C \text{ is bad}) = (1 - R_A R_{A'})(1 - R_B R_{B'})$$

where $(1 - R_A R_{A'})$ is the unreliability of the path in which A and A' are in series, and similarly $(1 - R_B R_{B'})$ is the unreliability of the second path in which B and B' are in series. Because the two paths are in parallel, the product of their unreliabilities gives the unreliability of the parallel combination.

We can now write the unreliability of the whole system as follows:

$$Q_s = (1 - R_{A'})(1 - R_{B'}) \cdot R_c + (1 - R_A R_{A'})(1 - R_B R_{B'}) \cdot Q_c$$

If components A, C, and B are equal and each has a failure rate of λ_1, and the components A' and B' are also equal and have a failure rate of λ_2 each, we obtain

$$Q_s = (1 - e^{-\lambda_2 t})^2 e^{-\lambda_1 t} + (1 - e^{-(\lambda_1+\lambda_2)t})^2(1 - e^{-\lambda_1 t})$$

and because $R_s = 1 - Q_s$ we obtain the reliability of the system,

$$R_s = 1 - (1 - e^{-\lambda_2 t})^2 e^{-\lambda_1 t} - (1 - e^{-(\lambda_1+\lambda_2)t})^2(1 - e^{-\lambda_1 t})$$

Assuming the numerical values of $\lambda_1 = 0.01$, $\lambda_2 = 0.001$, and $t = 10$ hours,

$$R_s = 1 - (1 - 0.99)^2\, 0.9 - (1 - 0.89)^2\, 0.1 = 0.9987$$

We shall now show the general character of Equation (13.3) by applying it to series and parallel connections for which we have already derived formulas in a different way.

If two components with reliabilities R_1 and R_2 operate in series, we choose one of them as component X—for instance, that which has the reliability R_2—and write

$$Q_s = (1 - R_1) \cdot R_2 + 1 \cdot (1 - R_2) = R_2 - R_1R_2 + 1 - R_2 = 1 - R_1R_2$$

The term $1 \cdot (1 - R_2)$ may need an explanation. The probability of system failure if the second component is bad equals unity because it is then certain that the system has already failed. We therefore multiply the probability that the second component is bad $(1 - R_2)$ with the probability of system failure given that the second component is bad, which equals unity, and so we obtain the term $1(1 - R_2)$. System reliability is then obviously

$$R_s = 1 - Q_s = R_1R_2$$

When we apply the theorem to two components in parallel we obtain

$$Q_s = 0 \cdot R_2 + (1 - R_1)(1 - R_2) = 1 - R_1 - R_2 + R_1R_2$$

The first term on the right, $0 \cdot R_2$, is the product of the probability of system failure given that the second component is good and of the probability R_2 that the second component is good. Obviously, if the second component is good, the parallel system cannot fail and therefore its probability of failure equals zero. The reliability of the system of two parallel components is then

$$R_s = R_1 + R_2 - R_1R_2$$

as we already know.

In complex reliability calculations it is often necessary to apply the probability theorem of Equation (13.3) in several steps. To show this technique, let us consider four components, 1, 2, 3, and 4 in parallel. Let the components have the reliabilities R_1, R_2, R_3, and R_4 and let us specify that at least one of the four components must operate. We proceed by evaluating first the unreliability Q_s of this system:

$$Q_s = Q_s(\text{if 1 is good}) \cdot R_1 + Q_s(\text{if 1 is bad}) \cdot Q_1$$

The first term on the right must be zero if component 1 is good, because if 1 is good the system cannot fail and Q_s(if 1 is good) must be zero. Therefore, we are left with $Q_s = Q_s$(if 1 is bad)$\cdot Q_1$. As the next step we must find Q_s(if 1 is bad). So we consider the three remaining components, 2, 3, and 4, pick out the second one and write

$$Q_s(\text{if 1 is bad}) = Q_s(\text{if 2 is good}) \cdot R_2 + Q_s(\text{if 2 is bad}) \ Q_2$$

$$= Q_s(\text{if 2 is bad}) \cdot Q_2$$

because the first term on the right is again zero. Thus, we now have $Q_s = Q_s$(if 2 is bad)$\cdot Q_2 \cdot Q_1$, by combining the results obtained so far. If we still do not know the answer for Q_s(if 2 is bad), we consider the last two components, 3 and 4, pick out component 3, and write

$$Q_s(\text{if 2 is bad}) = Q_s(\text{if 3 is good}) \cdot R_3 + Q_s(\text{if 3 is bad}) \cdot Q_3$$
$$= Q_s(\text{if 3 is bad}) \cdot Q_3 = Q_4 \cdot Q_3$$

Therefore the result for the system of four parallel components is

$$Q_s = Q_1 Q_2 Q_3 Q_4$$
$$R_s = 1 - Q_1 Q_2 Q_3 Q_4$$

But if we specify that at least three (which may be any three) of the four different components must operate to have the system work, the calculation becomes somewhat more complicated. We again start with the same first step as applied above:

$$Q_s = Q_s(\text{if 1 is good}) \cdot R_1 + Q_s(\text{if 1 is bad}) \cdot Q_1$$

The first term on the right does not become zero now, however, because with one component good the system can still be in a failed state. We require three components to operate for system operation, and therefore system failure occurs only if at least two components are in a failed state. Thus, to obtain Q_s(if 1 is good) we must evaluate the probability that of the remaining three components at least two fail. Thus, we pick out component 2 and write:

$$Q_s(\text{if 1 is good}) = Q_s(\text{if 2 is good}) \cdot R_2 + Q_s(\text{if 2 is bad}) \cdot Q_2$$
$$= (1 - R_3)(1 - R_4) \cdot R_2 + (1 - R_3 R_4) \cdot Q_2$$

The first term on the right means that if 2 is good (R_2), and because we said that 1 is good, both remaining components 3 and 4 must fail to give system failure. The second term on the right means that if 2 is bad (Q_2), and knowing that 1 is good, at least one of the two remaining components, i.e., either 3 or 4, must fail.

A similar analysis must be made for the term Q_s(if 1 is bad):

$$Q_s(\text{if 1 is bad}) = Q_s(\text{if 2 is good}) \cdot R_2 + Q_s(\text{if 2 is bad}) \cdot Q_2$$
$$= (1 - R_3 R_4) \cdot R_2 + 1 \cdot Q_2$$

because here we assume that 1 is bad. So we need one more component to fail if 2 is good, and if 2 is bad, system failure has already occurred because 1 is also bad. Therefore, Q_s(if 2 is bad) equals unity if 1 is also bad.

We can now write for the unreliability of the whole system

$$Q_s = Q_s(\text{if 1 is good}) \cdot R_1 + Q_s(\text{if 1 is bad}) \cdot Q_1$$
$$= [(1 - R_3)(1 - R_4)R_2 + (1 - R_3 R_4)Q_2] \cdot R_1$$
$$+ [(1 - R_3 R_4)R_2 + Q_2] \cdot Q_1$$
$$= 1 - R_1 R_2 R_3 - R_1 R_2 R_4 - R_1 R_3 R_4 - R_2 R_3 R_4 + 3 R_1 R_2 R_3 R_4$$

And the system reliability will be

$$R_s = R_1R_2R_3 + R_1R_2R_4 + R_1R_3R_4 + R_2R_3R_4 - 3R_1R_2R_3R_4$$

Of course, we could have obtained the same result by developing the term $(R_1 + Q_1)(R_2 + Q_2)(R_3 + Q_3)(R_4 + Q_4) = 1$ and proceeding as shown in Chapter 11, Equation (11.29), for three components. But here we went purposely through this exercise because situations occur where the binomial expansion will not help, as for instance, in the very simple example of Figure 13.1 and where in complex calculations the step-by-step application of this probability theorem becomes a matter of necessity. It is therefore important that those who must perform complicated reliability analyses become familiar with the application of Equation (13.3), which is an invaluable tool for the reliability analyst.

Chapter 14

COMPONENT MODES OF FAILURE
AND SYSTEM RELIABILITY

A MODIFIED VERSION of Equation (13.2) finds application when a component or a unit within a circuit is subject to two or more modes of failure and the probabilities of each of these modes of failure must be evaluated to compute the system's reliability.

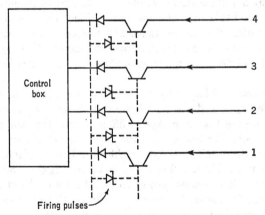

Fig. 14.1. A transistor switching circuit.

Let us consider the circuit shown in Figure 14.1. This circuit consists of four basic blocks, each having a transistor protected by a diode in the power circuit and by a zener in the firing circuit. The operation of this system requires that the transistors supply power to four elements in the control box in a fast sequence, 1-2-3-4-1-etc., so that when one transistor conducts, the preceding transistor is switched off. The switching sequence is regulated by the firing pulses. It is specified that system failure occurs if any of the four blocks fails short because the switching sequence is not maintained when a shorted block supplies power all the time. It is further specified that operation can continue if one block fails open so that only three blocks transmit power pulses in sequence, with one pulse always missing, but if two blocks fail open, the system is not capable of further operation. Thus, system reliability requires that no block fail short and not more than one block fail open. How do we calculate the reliability of this arrangement?

We apply Theorem (13.2) in the following way: We select the zener as component X because we assume that it is independent of the state of the other components, and write for the unreliability of one block, P(block failure) is equal to

P(block failure if zener is good) $\times$ P(zener is good)
$+$ P(block failure if zener has failed short)$\times$ P(zener fails short)
$+$ P(block failure if zener has failed open) $\times$ P(zener fails open)   (14.1)

This equation is identical to (13.2) except that, following Bayes' rule, we have decomposed the term $P(X$ is bad$)$ into the two modes in which X, the zener, can fail. We assume here that there are only these two modes in which the zener can fail. If there were a possibility of failure in modes other than open and short, we would have to add to Equation (14.1) a term which would take care of such probabilities. For instance, we could add P(block failure if zener has failed in other modes than open or short) $\times$ P(zener fails in other modes than open or short). This would exhaust all possibilities of failure. In our example we assume that this last term is zero.

Of course, one could ask why we make Equation (14.1) so complicated by introducing open and short probabilities. If we take a better look at Figure 14.1, we see that it makes a difference whether the components fail short or open. For instance, if the zener fails short, the firing pulse will not reach the transistor and the block will not switch. But if the zener fails open, the block will continue to switch. If the diode fails open, the block has failed because power cannot be transmitted to the box, but if the diode fails short the transistor can continue the switching and power will be supplied to the control box as required.

From Equation (14.1) we can calculate the unreliability of one block $Q_b = P$(block failure) if we specify precisely what is regarded as "failure." As to the reliability of the whole system of four blocks, the requirement was made that no block fail short and not more than one block fail open. If we designate the probability of one block's not failing short by R_{bs}, then the probability that no block out of four will fail short is R_{bs}^4, i.e., as if the four block were in series. And if we designate by R_{bo} the probability of one block's not failing open, then the probability that out of four blocks at least three will not fail open is $4R_{bo}^3 - 3R_{bo}^4$, according to the binomial expansion given in Equation (11.27).* The reliability of the system is then the probability that no block out of four will fail short and not more than one block out of four will fail open. Therefore, the reliability of the system is the product

$$R_s = R_{bs}^4 \cdot (4R_{bo}^3 - 3R_{bo}^4) \qquad (14.2)$$

assuming that "fail open" and "fail short" are the only two modes in which the blocks can fail—a good assumption.

We now have to compute the reliabilities R_{bs} and R_{bo}, i.e., the reliability that a block does not fail short, and the reliability that a block does not fail open. We compute first the unreliabilities Q_{bs} and Q_{bo} using Equation (14.1) and substituting for the term "block failure" the words "block fails short" when calculating Q_{bs}, and the words "block fails open" when calculating Q_{bo}.

Thus, the first term we must calculate is P(block fails short if zener is good) $\times P$(zener is good). To do this, we now select the diode as component X because it is independent of the state of the transistor, and thus P(block fails short if zener is good) is equal to

P(block fails short if diode is good) $\times P$(diode is good)
+ P(block fails short if diode has failed short) $\times P$(diode fails short)
+ P(block fails short if diode has failed open) $\times P$(diode fails open)

All these terms refer to the condition that the zener is good. Thus, P(block fails short if diode is good) equals the probability of the transistor's failing short when both the zener and the diode are good. But because the zener and the diode are to provide complete protection for the transistor, this unreliability will be either zero or an extremely small value, which means that the failure rate of the transistor can be assumed zero as long as the zener and the diode are good and rated temperatures are not exceeded. Next we have P(block fails short if diode has failed short). This amounts to the probability of the transistor failing short when the zener is good but the diode is shorted out and does not protect the transistor. In this condition the transistor operates at a higher stress level, i.e., it may be subject to higher electrical peaks, and therefore it may have a comparatively high failure rate. Finally, we have P(block fails short if diode has failed open), the zener still being considered as good. This probability amounts to zero, because with the diode failed open, the block will not fail short.

* $R^4 + 4R^3Q = R^4 + 4R^3(1 - R) = 4R^3 - 3R^4$

Thus, we have analyzed the term P(block fails short if zener is good) and have found that for all practical purposes this term amounts to the probability of the transistor's failing short when the zener is good but the diode is shorted out. We would analyze similarly, the term P(block fails short if the zener has failed short), but we assume here that this term will amount to zero because with a shorted zener the transistor will not trigger; also, in this particular case the maximum forward voltage is considerably less than the transistor's rated voltage, so the block will not fail short. The last term, then, is P(block fails short if zener has failed open). We analyze this term in the same way as we did the term, P(block fails short if the zener is good). Summing up and eliminating those terms which we can reasonably expect to be zero, we obtain for the probability of the block failing short (Q_{bs}) the following expression:

$$
\begin{aligned}
Q_{bs} = {}& [P(\text{transistor fails short if zener is good but diode has failed short}) \\
& \times P(\text{diode shorts}) \times P(\text{zener is good})] \\
& + [P(\text{transistor fails short if zener is open and diode is good}) \\
& \times P(\text{diode is good}) \times P(\text{zener opens})] \\
& + [P(\text{transistor fails short if zener is open and diode has failed short}) \\
& \times P(\text{diode shorts}) \times P(\text{zener opens})] \qquad (14.3)
\end{aligned}
$$

As is obvious, here we have broken down Q_{bs} into the probabilities of the individual components at the various stress levels. We now apply the same procedure to compute the probability of the block failing open (Q_{bo}) which amounts to

$$
\begin{aligned}
Q_{bo} = {}& [P(\text{transistor fails open if zener is good and diode is good}) \\
& \times P(\text{diode is good}) \times P(\text{zener is good})] \\
& + [P(\text{transistor fails open if zener is good but diode has failed short}) \\
& \times P(\text{diode shorts}) \times P(\text{zener is good})] \\
& + [1 \times P(\text{diode opens}) \times P(\text{zener is good})] \\
& + [1 \times P(\text{zener shorts})] \\
& + [P(\text{transistor fails open if zener has failed open and diode is good}) \\
& \times P(\text{diode is good}) \times P(\text{zener is open})] \\
& + [P(\text{transistor fails open if zener has failed open and diode has} \\
& \text{failed short}) \times P(\text{diode shorts}) \times P(\text{zener is open})] \\
& + [1 \times P(\text{diode opens}) \times P(\text{zener is open})] \qquad (14.4)
\end{aligned}
$$

To calculate the numerical value of this system's reliability, we must express the probabilities P of which Q_{bs} and Q_{bo} consist as functions of time, and we must know the rates at which the components fail in the various modes of failure. When there are only two modes of failure, opens and shorts, they may be mutually exclusive and we can assume that the component's failure rate λ is the sum of its rate failing short λ_s and its rate failing open λ_o; therefore

$$\lambda = \lambda_s + \lambda_o$$

If we further assume that the components are utilized only in their useful life period, the rates of failure will be constant for a given stress level. As stress levels change, the failure rates will also change. However, these changes are not linear. Small increases in the stress level may cause large increases in the failure rate, λ. Therefore, λ_s and λ_o will also be subject to changes, but they do not need to change proportionally. For instance, with an increasing stress level, the rate of failing short λ_s may be increasing slower than the rate of failing open λ_o, and from some stress level upwards only λ_o may be increasing, while λ_s may start to decrease. The actual ratio of λ_s to λ_o can be determined by tests. Changes of failure rates with changing stress levels are discussed in the following chapter.

In our example of four transistor blocks, we must consider several stress levels determined by whether the protective devices are good or have failed. The stress level to which the zener is exposed is determined by the firing pulses. We can assume that the distribution of electrical stresses is exponential and that the zener has therefore a constant overall failure rate λ_1, and constant rates of failing short and open, λ_{1s} and λ_{1o}, throughout the operation. The diode, however, may operate at two stress levels according to whether it is protected by the zener when the zener is good or is not protected when the zener has failed open. This is a conditional situation. The transistor may operate at four stress levels, i.e., when the zener and the diode are good, when the zener is good and the diode is shorted, when the zener is open and the diode is good, and finally, when the zener is open and the diode is shorted. We tabulate these failure rates as follows:

	Failure rate including all failure modes	*Rate of failing short*	*Rate of failing open*
Zener	λ_1	λ_{1s}	λ_{1o}
Diode not protected (zener open)	λ_2	λ_{2s}	λ_{2o}
Diode protected by zener	λ_3	λ_{3s}	λ_{3o}
Transistor not protected (zener open, diode short)	λ_4	λ_{4s}	λ_{4o}
Transistor protected by zener only (diode short)	λ_5	λ_{5s}	λ_{5o}
Transistor protected by diode only (zener open)	λ_6	λ_{6s}	λ_{6o}
Transistor protected by zener and diode	λ_7	λ_{7s}	λ_{7o}

Of course, it is impossible ever to get the exact numerical values of all these failure rates. However, some reasonable estimates can be made by combining available test results with assumptions based on engineering experience and judgment—which is of greatest importance in reliability work and cannot be replaced by abstract mathematical methods, or by an impractically large amount of testing. We shall see in the next chapter that it is possible to obtain a fairly good picture of how component failure rates change with stress level changes. Also, information can be obtained from experienced electronic engineers as to the approximate proportion of shorts to opens for certain semiconductor components. But this again depends on the circuit conditions such as the magnitude of transients, their duration, and the capability of components and heat sinks to dissipate the generated heat. In a certain application it may be found that about one-half of the zener failures are shorts and one-half are opens. For instance, if in the above example the failure rate of the zener is $\lambda_1 = 0.00002$, we would assume that $\lambda_{1o} = 0.00001$ and $\lambda_{1s} = 0.00001$. But for diodes and transistors a better assumption would be a $9:1$ ratio of shorts to opens, depending, of course, on the results of a steady-state and transient analysis of the entire electric circuit.

When the failure rate estimates are obtained, it becomes possible to express the probability equations (14.3) and (14.4) for the above example in a numerical form as exponential functions of the operating time t, and therefore also the probabilities of one transistor block's not failing short, R_{bs}, and not failing open, R_{bo}, as well as the over-all reliability of the system of four blocks as the probability that no block fails short and not more than one block fails open.

The transformation of the probability equations into equations containing exponential functions is quite an intricate mathematical procedure when changes of the failure rates with changing stress levels are considered. The intricacy lies in the fact that the times at which failure rate changes occur in the diode and in the transistor depend on the variable times t_1 and t_2 at which the zener may fail open or short; the times at which the transistor failure rates change also depend on the variable times at which the state of the diode changes. Here we have to do with dependent events of a similar nature as discussed in Chapter 12 in connection with stand-by components where we first had to evaluate the failure density functions to obtain the probabilities of failure Q. In the following pages we shall show the principles of the exact procedures which also apply to this example.

The calculations can be simplified considerably if we assume that the failure rates of the components do not change. For instance, we assign to the zener the fail-open and fail-short rates λ_{zo} and λ_{zs}, to the diode λ_{do} and λ_{ds}, and to the transistor λ_{to} and λ_{ts}. We assume thus that the transistor will have these failure rates regardless of the state of the diode and the zener, and the diode regardless of the state of the zener.

For (14.3) we then write

$$Q_{bs} = q_{ts}q_{ds}r_z + q_{ts}q_{zo}r_d + q_{ts}q_{ds}q_{zo} \qquad (14.5)$$

and for (14.4)*

$$Q_{bo} = q_{to}r_dr_z + q_{to}q_{ds}r_z + q_{do}r_z + q_{zs}$$
$$+ q_{to}q_{zo}r_d + q_{to}q_{ds}q_{zo} + q_{do}q_{zo} \qquad (14.6)$$

where q_o, q_s and r are the fail open, fail short and survival probabilities of the various components, as shown on page 138. For instance, in the case of the diode we have $q_{do} = \dfrac{\lambda_{do}}{\lambda_d}(1 - e^{-\lambda_d t})$, $q_{ds} = \dfrac{\lambda_{ds}}{\lambda_d}(1 - e^{-\lambda_d t})$ and $r_d = e^{-\lambda_d t}$, where $\lambda_l = \lambda_{lo} + \lambda_{ds}$.

These equations can be readily evaluated as can, therefore, Equation (14.2) for the system's reliability. If the selected nonchanging failure rates are good estimates of the average values, the final result will be quite realistic. In practical reliability work, except when the utmost precision is required, we follow this simplified procedure when complex networks and systems are being analyzed.

Whether the simplified or the exact procedure is used, the equation for system reliability remains the same, i.e., $R_s = R_{bs}^4(4R_{bo}^3 - 3R_{bo}^4)$. An inspection of this equation immediately shows that the term in the brackets, which is the system's reliability of not failing open, is much closer to unity than the term R_{bs}^4, which is the system's probability of not failing short. Therefore, efforts to make the system more reliable would concentrate in the first place on reducing the system's probability to fail short.

We shall now look into the problem of system reliability calculations when the failure rates of components depend on what happens to other components. A few simple cases will be discussed.

As a first step we re-examine the general case of parallel operation of components derived in Chapter 11. We assumed there that when a parallel component fails, the other surviving components will maintain the same failure rate. Thus, for two equal components in parallel we had the equation $R = 2e^{-\lambda t} - e^{-2\lambda t}$, which implies that each of the two components has a failure rate λ while they operate in parallel, and when one fails, the other component being of an adequate rating will continue to perform the required function alone, maintaining the same failure rate λ.

---

* We would assume the first term on the right to be zero as we did in (14.3) because the transistor is fully protected.

This is true in many operations where degradation of performance is accepted and taken into account. For instance, a system of two 40-kva aircraft generators supplies 80 kva full capacity, but when one generator fails the other will supply only 40 kva because nonessential loads are immediately disconnected to avoid the tripping of the surviving generator. Thus, the failure of one generator does not cause the electrical system to fail to supply the essential 40 kva, and the surviving generator operates essentially at the same stress level as before when both were operating. Thus, from the point of view of the system's function to supply under all conditions an essential power of 40 kva, the two generators are operating in parallel and at the same stress level—which means with the same failure rate, even after one of them fails. And from the point of view of the system's capability to supply the full 80 kva, the two generators operate in series, because if one fails the 80 kva are no longer supplied, and this constitutes system failure to perform its other function, i.e., that of supplying 80 kva. The system thus performs two functions—for one of them it is a series system and for the other it is a parallel system, and the system reliabilities of performing these two functions are calculated accordingly by means of the previously derived equations for series and parallel operation.

But the situation changes when two parallel components share the

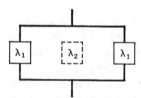

Fig. 14.2. Two load-sharing components in parallel.

full load 50-50, and when one fails the other must carry 100 per cent of the full load, being rated for such operation. When both components have a failure rate of λ_1 each while operating in parallel, the surviving component will assume a failure rate of λ_2 after the other component's failure because the stresses acting on it will increase. Figure 14.2 shows the block diagram of such an arrangement.

To calculate the reliability of this parallel arrangement of components, we can imagine that there exists in stand-by a third hypothetical component with a failure rate λ_2, and if any one of the two operating components fails, the parallel system is out of action and the stand-by component takes over to continue the operation alone. The probability that any one of the two parallel components fails is from this viewpoint a series case, and we may consider the two operating components as a single series unit which has a failure rate of $2\lambda_1$ and therefore a reliability of $e^{-2\lambda_1 t}$. The whole system can be thought of as this unit backed up by a stand-by component which has a failure rate λ_2 from the moment when it takes over the operation, i.e., from the moment when one of the operating components fails. Thus, Figure 14.2 can be redrawn into a block diagram as shown in Figure 14.3. The reliability of a system of two equal components in parallel which change failure rates from λ_1 to λ_2 if one fails is thus equivalent to the case of two stand-by components with different failure rates, the operating component having a failure rate of $2\lambda_1$ and the stand-by component a failure rate of λ_2. Therefore, Equation (12.13) can immediately be applied and the reliability of the system of two parallel components is

$$ R = e^{-2\lambda_1 t} + \frac{2\lambda_1}{\lambda_2 - 2\lambda_1} \left(e^{-2\lambda_1 t} - e^{-\lambda_2 t} \right) \tag{14.7} $$

When the two components in parallel are not equal, the problem becomes more complicated. Let the two components, while both operate in parallel, have the failure rates λ_A and λ_B, but if component B fails first, A will change its failure rate from λ_A to λ_A', and if A fails first, B will change its failure rate from λ_B to λ_B'. We can thus imagine that we have two hypothetical components with failure rates λ_A' and λ_B' waiting in stand-by.

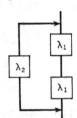

Again, the parallel system falls out of action if any of the two operating components fails, and we therefore consider the parallel components as a series unit with a failure rate of $\lambda_A + \lambda_B$. Now if A fails first, the stand-by component with a failure rate of λ_B' takes over to continue the operation, but if B fails first, the other stand-by component

Fig. 14.3. The equivalent reliability diagram.

with a failure rate of λ_A' takes over. Thus the stand-by system can fail in two ways:

1. Component A fails at time t_A, B takes over with a failure rate λ_B' and fails at or prior to time t.
2. Component B fails at time t_B, A takes over with a failure rate λ_A' and fails at or prior to time t.

These two events have the probabilities $P(1)$ and $P(2)$ which can be calculated by forming their density functions as shown in Chapter 12 and by subsequent integration. The system's unreliability Q_s is then the sum of these two probabilities, i.e.,

$$ Q_s = P(1) + P(2) $$

and system reliability is

$$ R_s = 1 - Q_s $$

The calculation is as follows:*

* This calculation and the resulting equations were supplied by Mrs. Julie Foster, mathematician at the Boeing Airplane Company, Renton, Washington.

$P(1) = P(A$ fails by t, given that B works, and B fails by t, given that A has failed at t_A)

$$= \int_{t_A=0}^{t} \lambda_A e^{-\lambda_A t_A} e^{-\lambda_B t_A} \int_{t_B=t_A}^{t} \lambda_B' e^{-\lambda_B'(t_B-t_A)}\, dt_B\, dt_A$$

$$= \frac{\lambda_A}{\lambda_A + \lambda_B}(1 - e^{-(\lambda_A+\lambda_B)t}) + \frac{\lambda_A}{\lambda_A + \lambda_B - \lambda_B'}(e^{-(\lambda_A+\lambda_B)t}$$
$$- e^{-\lambda_B't}) \quad (14.8)$$

$P(2) = P(B$ fails by t, given that A works, and A fails by t, given that B has failed at t_B)

$$= \int_{t_B=0}^{t} \lambda_B e^{-\lambda_B t_B} e^{-\lambda_A t_B} \int_{t_A=t_B}^{t} \lambda_A' e^{-\lambda_A'(t_A-t_B)}\, dt_A\, dt_B$$

$$= \frac{\lambda_B}{\lambda_A + \lambda_B}(1 - e^{-(\lambda_A+\lambda_B)t}) + \frac{\lambda_B}{\lambda_A + \lambda_B - \lambda_A'}(e^{-(\lambda_A+\lambda_B)t}$$
$$- e^{-\lambda_A't}) \quad (14.9)$$

The reliability of this stand-by system is then 1 minus the sum of $P(1) + P(2)$:

$$R_s(t) = e^{-(\lambda_A+\lambda_B)t}\left[1 - \frac{\lambda_A}{\lambda_A + \lambda_B - \lambda_B'} - \frac{\lambda_B}{\lambda_A + \lambda_B - \lambda_A'}\right]$$
$$+ \frac{\lambda_A e^{-\lambda_B't}}{\lambda_A + \lambda_B - \lambda_B'} + \frac{\lambda_B e^{-\lambda_A't}}{\lambda_A + \lambda_B - \lambda_A'} \quad (14.10)$$

If $\lambda_B' = \lambda_A + \lambda_B$:

$$P(1) = \int_{t_A=0}^{t} \lambda_A e^{-\lambda_A t_A} e^{-\lambda_B t_A} \int_{t_B=t_A}^{t} (\lambda_A + \lambda_B)e^{-(\lambda_A+\lambda_B)(t_B-t_A)}\, dt_B\, dt_A$$

$$= \frac{\lambda_A}{\lambda_A + \lambda_B}(1 - e^{-(\lambda_A+\lambda_B)t}) - \lambda_A t e^{-(\lambda_A+\lambda_B)t} \quad (14.11)$$

Similarly, if $\lambda_A' = \lambda_A + \lambda_B$,

$$P(2) = \frac{\lambda_B}{\lambda_A + \lambda_B}(1 - e^{-(\lambda_A+\lambda_B)t}) - \lambda_B t e^{-(\lambda_A+\lambda_B)t} \quad (14.12)$$

And if $\lambda_A' = \lambda_B' = \lambda_A + \lambda_B$,

$$R_s(t) = e^{-(\lambda_A+\lambda_B)t} + (\lambda_A + \lambda_B)t e^{-(\lambda_A+\lambda_B)t} \quad (14.13)$$

which can also be obtained from Equation (12.3).

Equations (14.7) and (14.10) reduce to the simple parallel reliability equations (11.16) and (10.11) respectively when we assume that the components do not change their failure rates.

If three or more components share the load in parallel operation, but each of them is capable of carrying the full load alone if all other components fail, the surviving components assume higher and higher failure rates as their number is reduced because of other failing components. If the components are unequal, the same procedures will be used as outlined above, although the equations necessarily become more complex. On the other hand, if the components are equal, as is usually the case in parallel operations, the procedure will be similar to that used for two equal components earlier in this chapter and illustrated by Figures 14.2 and 14.3.

We imagine a unit consisting of four parallel components with a unit series failure rate $4\lambda_1$, backed up by a stand-by unit of three components with unit failure rate $3\lambda_2$, this again backed up by a unit of two components with $2\lambda_3$, and finally, this backed up by the last stand-by component which has a failure rate λ_4. We then use Equation (12.19) to solve this problem, substituting the unit failure rates for component failure rates.

Besides the mathematical complexity of this rigorous treatment of parallel operating components which change their failure rates, there is also the complication that, for each component in such system, several failure rates must be considered. When this becomes too cumbersome, simplified comparative analyses of parallel systems are made instead, using the methods outlined in Chapter 11.

However, the above discussion shows that in load-sharing parallel arrangements where components change their failure rates, higher reliabilities can be achieved than in stand-by operations, where the same components are subject to higher operational stresses all the time. In addition, stand-by operations usually involve the use of sensing-switching devices which further reduce reliability. Of course, the final choice of arrangement depends on the performance requirements and whether a component is suited for parallel or stand-by operation. We have already mentioned that many components, such as resistors and capacitors, cannot always be used in parallel reliability operations. Thus, each system, parallel and stand-by, has its merits with proper application. Where one is used, the other will probably not fit in at all because of engineering considerations. And when a choice is possible, it is usually enough to know the mean time between failures of the two systems which have to operate either in parallel or in stand-by to increase reliability. For instance, it is more reliable to have two transmitter-receivers in stand-by than in parallel, especially when long space flights are involved and also where wear-out plays a role.

Let the one transmitter have a mean time between failures of m_1 and the other m_2. In stand-by their combined mean time between failures will be $m_1 + m_2$, and in parallel it will be $m_1 + m_2 - m_1m_2/(m_1 + m_2)$ as shown in Equation (11.14), because the stress level on the one transmitter will not change if the other fails. If an unmanned flight is considered, then in a stand-by arrangement the second transmitter must be switched on if the first fails, but this switching is a one-time, one-cycle operation which takes only fractions of a second and is triggered from the ground so that no failure-sensing circuitry is necessary in the space vehicle. Such a one-time switching operation can have a very high reliability. And if we were to assume that the two transmitters can fail only when they wear out, each having the same mean wearout life M, the expected life of the stand-by system will be twice that of the parallel system, if the standard deviation of lives is small. Also, if we had one operating transmitter and two in successive stand-by, the expected life would be $3M$ as

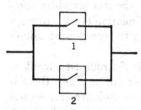

against only about one-third of it if the three transmitters were to operate simultaneously, i.e., in parallel.

To complete this chapter on components with several modes of failure and on changing failure rates, we shall discuss the case of parallel operating switches.

Fig. 14.4. Two switches in parallel.

Let λ_o and λ_s be the constant fail-open and fail-short rates of a switch, so that $\lambda = \lambda_o + \lambda_s$. Its probability of failing open by the time t, given that it has not failed short, is then $q_o = \dfrac{\lambda_o}{\lambda}(1 - e^{-\lambda t})$. Similarly, we have $q_s = \dfrac{\lambda_s}{\lambda}(1 - e^{-\lambda t})$. The probability that the switch will fail in either mode, open or short, is therefore

$$q = q_o + q_s = \frac{\lambda_o + \lambda_s}{\lambda}(1 - e^{-\lambda t}) = 1 - e^{-\lambda t} \qquad (14.14)$$

and its probability that it will fail neither open nor short is simply $r = e^{-\lambda t}$.

To calculate the reliability R of two switches operating simultaneously in parallel so that if one switch fails the other is capable of performing the switching function alone, we use Bayes' theorem for the block diagram shown in Figure 14.4.

$$P(\text{system failure}) = P(\text{system failure if 2 is open})P(2 \text{ is open})$$
$$+ P(\text{system failure if 2 is short})P(2 \text{ is short})$$
$$+ P(\text{system failure if 2 is good})P(2 \text{ is good}) \quad (14.15)$$

This amounts to the parallel system's unreliability Q_p. Transcribed into an exponential form, and when the two switches have equal failure rates, we have

$$Q_p = qq_o + q_s + q_s r = q + r(q_s - q_o) = q + rq \frac{\lambda_s - \lambda_o}{\lambda} \quad (14.16)$$

and the reliability R_p of the parallel switch system is then

$$R_p = r + r(q_o - q_s) = r + rq \frac{\lambda_o - \lambda_s}{\lambda} \quad (14.17)$$

where r is the reliability that a switch will fail neither open nor short, λ_o is its fail-open rate, and λ_s its fail-short rate. In the above model we have neglected the possibility that the switches may change their rates of failure when they operate alone after the other switch has failed. Therefore, this model is good for such circuits where the switches are loaded only by a small fraction of their actual rating so that it makes practically no difference in stress levels whether two operate in parallel or one is left operating alone.

Equation (14.17) shows that paralleling two switches would increase reliability only if $q_s < q_o$, and therefore, if $\lambda_s < \lambda_o$. If $\lambda_s = \lambda_o$, the reliability of two switches in parallel would equal the reliability of a single switch, but maintenance would be doubled. And if $\lambda_o < \lambda_s$, two switches in parallel would be a less reliable arrangement than a single switch. Unfortunately, some

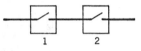

Fig. 14.5. Two switches in series.

switches (such as semiconductor switches) have a definite tendency to fail short rather than open. The ratio of $\lambda_s : \lambda_o$ may be as high as 10:1 or even higher. If this is the case, and a single switch is not reliable enough to perform a required switching function with a very high reliability, we would put two switches in series as shown in Figure 14.5. Applying again the probability equation (14.15), we obtain for the unreliability of a series arrangement of two switches,

$$Q_s = q + rq \frac{\lambda_o - \lambda_s}{\lambda} \quad (14.18)$$

and for the reliability R_s,

$$R_s = r + rq \frac{\lambda_s - \lambda_o}{\lambda} \quad (14.19)$$

Thus, we see that the series arrangement improves the reliability of switching if $\lambda_o < \lambda_s$.

The above conclusions are correct only in cases where failure rates do not change at all or change only insignificantly. The situation may become quite different when a switch, when operating alone, is stressed to its full current-carrying capacity I, yet switches only $I/2$ when two operate in parallel. The rates of failure under the full-current condition may be much larger than under the half-current condition. Assume that under full current the switch has the rates of failure λ_o' and λ_s', and under the half-current condition (i.e., when operating in parallel) λ_o and λ_s.

To solve this problem we proceed in a way similar to the derivation of Equation (14.7) which gives the reliability of two components in parallel which change their failure rates if one fails.

Thus, we imagine that the two parallel operating switches form an operating unit, and that there is a third hypothetical switch with the rates of failure λ_o' and λ_s' which takes over the operation if any of the two switches in the operating unit fails open, which also constitutes a failure of the entire unit so that the hypothetical switch then operates alone. System failure occurs if (a) any of the two switches in the unit fails short, or (b) any of the two switches in the unit fails open and the hypothetical third switch which is in stand-by also fails, open or short. Therefore, there are two probabilities of system failure:

$$Q_a = \frac{\lambda_s}{\lambda} (1 - e^{-2\lambda t})$$

$$Q_b = \frac{\lambda_o}{\lambda} (1 - e^{-2\lambda t}) + \frac{2\lambda_o}{2\lambda - \lambda'} (e^{-2\lambda t} - e^{-\lambda' t})$$

And because the events (a) and (b) cannot occur simultaneously, the unreliability of the parallel system amounts to

$$Q_p = Q_a + Q_b$$

and system reliability is

$$R_p = 1 - Q_a - Q_b = e^{-2\lambda t} + \frac{2\lambda_o}{2\lambda - \lambda'} (e^{-\lambda' t} - e^{-2\lambda t}) \qquad (14.20)$$

The expression for Q_b is the probability of failure of a system of two components, each having a failure rate of $\lambda = \lambda_o + \lambda_s$, and which together have a series fail-open rate of $2\lambda_o$. When one of the components fails open while the other works, the surviving component continues to operate at a failure rate of $\lambda' = \lambda_o' + \lambda_s'$.

If the switches were to fail short only, so that $\lambda_o = 0$, then Equation (14.20) reduces to $R_p = e^{-2\lambda_s t}$, which means that in this case it is less reliable to use two switches in parallel than to have a single switch with $R = e^{-\lambda_s t}$, because from the probability point of view two paralleled switches which only fail short appear as being in series. If either of them fails, i.e., fails short, the system fails, and if they are in series circuitwise, so that they carry the same current whether both operate in series or the one fails short and the other continues switching alone, and assuming that this does not change the failure rate of the surviving switch, the reliability of this system becomes $2e^{-\lambda_s t} - e^{-2\lambda_s t}$. Therefore, a system consisting of two switches operating circuitwise in series where the switches can only fail short, from the reliability point of view, appears as a parallel reliability system.

The above consideration of parallel and series operating switches, with and without failure rate changes, applies equally well to moving contacts when the contact is considered as a separate component. Similar considerations apply when switches or contacts are required to close or to open a circuit. It is obvious that when a circuit is originally open and should be closed, it is more reliable to have two contacts or switches parallel to do this job. But if the requirement is to open an originally closed circuit, then the series arrangement is more reliable.

Chapter 15

COMPONENT FAILURE RATES AT SYSTEM STRESS LEVELS

THE METHODS AND TECHNIQUES we have developed so far for calculating the reliability of various combinations of components enable us to calculate the reliability of even the most complicated and complex systems. We have seen that components can combine in several ways to form a system. There is the series case, the parallel case, and the stand-by case. But there also exist situations which do not fit into any of these categories and Bayes' theorem must be used.

All system reliability calculations have one thing in common which is fundamental to the probability calculus: The reliabilities of the basic components which form the system must be known. These component reliabilities must be determined from trials, observations, or tests which yield basic probability information in the form of failure rates, mean times between failures, or mean number of cycles between failures. We shall learn later how these basic parameters are measured or estimated by means of reliability tests.

In the discussion of failure distributions we have seen that component failure rates may be constant or may be functions of the component age. Accordingly, the reliability of components also may be independent of or may depend on age. Thus, reliability, although always a function of the operating time t, does not need to be a function of the component's or system's age T. When reliability is independent of the age, a component or a system which has a reliability $R(t)$ for an operating interval t will have the same reliability $R(t)$ for a later operating interval of the same length t, even if in the meantime the component or the system has been in operation for hundreds or thousands of hours. This behavior is characteristic of exponential components which do not deteriorate with time and are subject to chance failures only. Systems built with such components can be exponential if the components combine in series, but will not be exponential if the components combine in parallel or in any way other than in series. Still, whether exponential or not, all systems composed exclusively of exponential components will have age-independent reliabilities if no component in the system fails and if any failed components in the system are replaced by equally good ones. The system reliability will again be the same for a later operation of the same length t. Therefore, system reliabilities can be kept age-independent almost indefinitely if only exponential components are used in the system and if failed components are replaced by good ones. On the other hand, component reliabilities cannot be kept indefinitely age-independent. Sooner or later components begin to wear out, and therefore they should be replaced before this happens, even if they have not yet failed.

System failures are caused by component failures. When components can fail only because of chance, the system will fail only because of chance. But if components are allowed to wear out in a system—which means that no preventive replacements are made and components are replaced only as and when they fail—such a system comes into a state of wearout with a very high failure rate and low reliability. If we consider chance only, component mean times between failures in conservatively designed systems provided with adequate safety factors can reach hundreds of thousands, even millions of hours. As we know, this does not mean that a component could really operate for such a long period, and it is most probable that the component will wear out long before that time. Wearout lives of components are normally much shorter than their mean times between failures. As we have said, 10,000 hours is a very good average life. Thus, a component which has a mean time between failures of, say, 100,000 hours within the first few thousand hours of its life will have in a system which is designed to operate for years a mean time between failures of only, say, 10,000 hours or less if preventive replacement is not applied. Systems in which components are not preventively replaced therefore reach a wearout state with greatly increased component failure rates.

In reliability engineering we are concerned with making systems as reliable as possible or as practical. Therefore, wherever reliability is of importance, we endeavor to schedule proper replacement times for the components so as to be sure that we operate components only within their useful life period. This policy is of the utmost importance for systems which must be used for extended periods of time and where regular maintenance can be provided. It does not apply to short-life systems such as missiles, where the probabilities of component failures are usually governed by chance, and wearout probabilities are extremely low. Thus, systems engineered and maintained with due consideration to reliability will fail only of chance, and that means that in all reliability analyses of such systems the exponential formula $e^{-\lambda t}$, where λ stands for the chance failure rate, can be safely used for all components. This greatly simplifies reliability calculations of complex systems.

Complex systems consist of a multitude of components combined in various ways. When the system is in operation, some of the components will operate all the time, but some will operate only for short stretches, or only for a few cycles, and some components may perform only a single short operation when the system is started or shut off. Therefore, during a given system operating time t for which we want to calculate the system's reliability, $R_s(t)$, only a certain number of components operates for the same length of time, while other components operate for shorter times only. Now we know that the reliability of a component is a function of the operating time of that component; it is the probability that the component will not fail in a certain time interval of its operation. Thus, when we have a system in which not all components operate for the same length of time, we must coordinate the components' operating times to the system's operating time in order to calculate the system's reliability.

For instance, we want to calculate the reliability of a system for 10 hours of system operation. Knowing that system reliability is composed of the reliabilities of its components, we must relate these reliabilities to the system's operating time. The reliabilities of those components which operate all the time in the system will be expressed for 10 hours of component operation. Thus, if one of these components has a failure rate λ, its reliability for 10 hours of system operation will be $\exp(-10\lambda)$. But if a component operates only 1 hour during a system operation of 10 hours, its reliability for 10 hours of system operation will be $\exp(-1\lambda)$, which is its reliability for 1 hour of component operation. And if another component operates only for 5 minutes in these 10 hours of system operation, as might be the case with components concerned only with starting up the system, the reliability of this component will be $\exp(-1\lambda/12)$ for the entire 10 hours of system operation. Thus, for a given system operating time, we consider only the "energized" time of the individual components. We do this on the assumption that during the periods when the component is not energized, i.e., when it is not in operation, its failure rate is zero.

The reliabilities of components which operate for various lengths of time in a system may still combine in series, and they usually do. Thus the product rule of reliabilities still applies, and the series system reliability $R_s(t)$ for t hours of system operation is then the product of the component reliabilities:

$$R_s(t) = \Pi\ R_i(t_i) \qquad (15.1)$$

where $R_i(t_i)$ is the reliability of the ith component which during t system hours is energized for t_i hours. Expressed in terms of the exponential functions of the component reliabilities, Equation (15.1) can be written as

$$R_s(t) = e^{-\lambda_1 t_1} e^{-\lambda_2 t_2} \cdots e^{-\lambda_i t_i} \cdots e^{-\lambda_n t_n}$$

$$= e^{-(\lambda_1 t_1 + \lambda_2 t_2 + \cdots + \lambda_i t_i + \cdots + \lambda_n t_n)}$$

$$= \exp\left(-\sum_{i=1}^{n} \lambda_i t_i\right) \qquad (15.2)$$

When we compare this equation with Equation (10.14) in Chapter 10, we see that the operating times of the components are not equal now, whereas in Equation (10.14) all component operating times were t. In Equation (15.2) it is essential that the component times t_1, t_2, etc., be the energized times of the components within the system's operating time t.

We can convert Equation (15.2) to the common system time basis t by introducing the concept of component *duty cycle*. We define the duty cycle of the ith component as $d_i = t_i/t$, which is the ratio of the component's operating time or energized time t_i to the system's operating time t. Thus, the component is de-energized for the period $t - t_i$, and is energized for the period $t_i = d_i t$. Equation (15.2) then becomes

$$R_s(t) = \exp\left(-\sum_{i=1}^{n} \lambda_i d_i t\right) \qquad (15.3)$$

The reliability of the ith component which operates in the system at a duty cycle of d_i is thus $\exp(-\lambda_i d_i t)$ for a system operating time t. The practical meaning of this is that a component known to have a failure rate of λ_i per 1 hour of component operating time assumes in the system a failure rate of $\lambda = \lambda_i d_i$ per 1 hour of system operation, because it operates at a duty cycle d_i. This procedure is also used when several subsystems must cooperate in a system to achieve system success, but the subsystems, or some of them, do not operate full time in the system time.*

A similar case arises when two systems must operate in succession to achieve a required function. For instance, the success of a missile depends on its launching and the timing thereof—not only on the reliability of the missile itself after launching. Obviously, the launching system and the missile have different operating times t_l and t_m. But here the time t_l is not a portion of the missile system time t_m. Still, the probability of success is the product of the reliabilities of the launching system and of the missile:

$$R = R(t_l) \times R(t_m)$$

If the launching system has a series failure rate of λ_l and the missile has a failure rate of λ_m, we can write $R = \exp(-\lambda_l t_l - \lambda_m t_m)$. The launching

---

* The case where systems contain components whose failure rates are expressed in terms of "failures per one operating cycle" is described in Chapter 10.

system fails on the average once every $1/\lambda_l$ hours of launching system operating time and the missile fails once every $1/\lambda_m$ hours of missile operating time. What is the mean time between failures of the combined operation?

We use the combined time $t = t_l + t_m$ as combined system reference time. In this time t the launching system operates at a duty cycle $d_l = t_l/(t_l + t_m)$, and the missile at a duty cycle $d_m = t_m/(t_l + t_m)$. Therefore the over-all failure rate for the combined operation is

$$\lambda = d_l\lambda_l + d_m\lambda_m = \frac{\lambda_l t_l + \lambda_m t_m}{t}$$

and the mean time between failures is $t/(\lambda_l t_l + \lambda_m t_m)$.

Another factor we must consider when analyzing the reliability of a system is the stress level at which components operate in the system, because this determines the failure rates the components will assume in the system. When we say that the failure rate of some component is constant, this refers to a particular operating and environmental stress level. The failure rates of components change significantly even with small changes of the stress level to which the components are exposed. It is the interaction between the strength of a component on one side, and the stress level or stress spectrum at which the component operates on the other side, which determines the failure rate of a component in a given situation. Thus, at different stress levels components necessarily assume different failure rates.

The physical process which determines component failure rate can be graphically illustrated as shown in Figure 15.1. The zigzag line illustrates the stress spectrum. Stress excursions occur at random time intervals. The solid line represents a component of strength S_1. We see here that in the operating interval from time $t = 0$ to time $t = T$ the stress level did not exceed the component strength, and therefore the component did not fail. But if a weaker component of strength S_2 (dashed line) were used, such a component would have already failed at a time $t = T_1$ and may have caused system failure. Thus, it would have had to be replaced. Actually, in the operating interval shown and when using components of strength S_2, we would have observed about eight failures, which means we would have made eight replacements. It also follows from the graph that if the stress peaks were to increase only slightly, component S_1 would also fail and two or more replacements would be necessary in the operating interval shown.

A further complication results because components taken from a production lot are never exactly of the same strength. Component strength in a lot can be imagined as normally or log normally distributed about a mean strength S_m with a certain standard deviation, which is then also the frequency distribution of each component in the lot. Similarly, we can also imagine the stress spectrum in the form of a frequency distribution about a mean stress M with some standard deviation. We can then illustrate the situation by the graph shown in Figure 15.2. The failure rate of a component stemming from a lot with the mean strength S_m and exposed to a stress spectrum with a mean stress M would be derived from

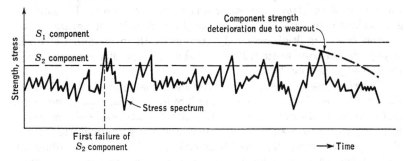

Fig. 15.1. Component strength vs. stress spectrum.

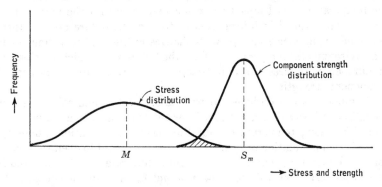

Fig. 15.2. Distribution of stresses and component strength.

the intersection of the two distributions. Of course, the standard deviations of the two distributions must also be known.

However, this approach is not practical because component strength and the stress spectrum cannot be defined in a satisfactory manner since both are a heterogenous mixture of a multitude of physical and other parameters. Component "strength" means not only its mechanical resistance against vibration, shock, pressure, or acceleration; it also includes thermal strength as a resistance against environmental and self-generated temperature and against temperature cycling; electrical strength, which includes the capability to withstand electrical potentials and changes of the potential such as frequency; resistance against humidity, corrosion, radiation, etc. Thus it is obvious that component strength cannot be expressed by some meaningful numerical value and that no units of strength which would include all these parameters can exist. The same deliberations apply to what we have called the "operational stress spectrum." There is a multitude of various stresses acting on the component in operation and no known method exists to express the sum of these stresses by some single numerical value. In reality we are dealing here with an example of the variability of nature in connection with a multidimensional, stress-strength space, and no amount of physical measurements will ever enable us to predict component failure rates by such methods.

Therefore, we must apply statistics and measure by statistical testing the observable effect of component failure in the time domain instead of trying to derive the exact spectra. We submit a population of components to an operating stress level and count how many components fail in a given operating time. This yields an estimate of component failure rate for that particular stress level. We can further change some of the stress parameters one by one—for instance, increase or decrease the voltage or the temperature—and repeat the whole experiment for the various conditions. This will yield information as to how the failure rates of these particular components change with changes in the applied voltage or in the environmental temperature. Thus the failure rate, which we measure in number of failures per unit time, is a substitute for a measure of component strength.

Components are designed to withstand certain nominal stresses in operation. We say they are "rated" for a definite operating voltage, current, frequency, temperature, vibration, shock, acceleration, humidity, etc. When a population of such components is exposed to operation under the rated operating conditions, a certain failure rate may be observed in statistical tests. We call this the *nominal* failure rate of components, because it refers to operation at rated values.

It is well known that when we increase the operational stresses or some of these stresses above the rated level, the component failure rate increases rather rapidly above the nominal failure rate. Conversely, the failure rate decreases when the stresses are decreased below the rated level. The graph in Figure 15.3 shows the general pattern of failure rate curves of electrical and electronic components.

The failure rate scale on the graph is logarithmically ascending. The graph indicates how to reduce component failure rates below their nominal values. For instance, if a component is to operate at its rated wattage (curve 1.0), a reduction of environmental stresses, such as a reduction of temperature by providing heat sinks and if necessary adding forced cooling, will immediately reduce the component's failure rate. In cases where an appreciable reduction of temperature means adding too much weight to the equipment, a better way may be to use a component of double rating and thus to operate it at only 0.5 of its rated wattage value. This

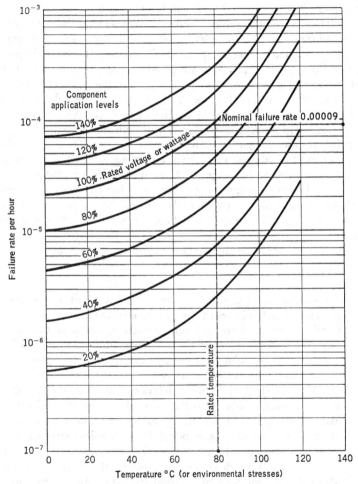

Fig. 15.3. Failure rate derating curves.

may be cheaper and may add less weight than, for instance, forced cooling. The graph shows that very appreciable gains in failure rate reduction are achieved when higher rated components are operated at one-half or even one-third of their rated values. A very good method for achieving high component reliabilities is to combine the two effects, i.e., reduce temperature and use components of higher ratings. This is a golden rule for designing reliable equipment, especially in electronic systems where component mean times between failures may be increased from an order of magnitude of 10^5 hours at rated operation to an order of magnitude of 10^6 and 10^7 hours, i.e., 10,000,000 hours and more at derated operation.* Thus, by derating components—using them at considerably lower than rated stress levels—the reliabilities of basically reliable components can be spectacularly further uprated. Of course, this very important reliability design principle does not relieve us of the necessity of using good debugging procedures so as to weed out the initially weak components which cause early failures. However, this is not a designer's task but belongs with the activities of quality control, receiving inspection of components, and functional testing of the built equipment (debugging) before it is released for service. The designer only specifies the procedures. But no amount of control, inspection, or testing can help if the designer has not paid attention to designing the equipment to be reliable. All that these controls and tests can do is to see to it that the built equipment is as reliable as designed and that early failures (weak components, poor connections, assembly errors, etc.) do not creep in or are eliminated if they do.

The stresses acting on components can generally be grouped into two categories:

1. Environmental stresses which are present whether a component in a system is actively operating or is in a state of quiescence, such as humidity, atmospheric pressure, radiation, chemical content, and impurities of the atmosphere, microbes, and environmental temperature at the place where the component is located. These stresses normally do not cause chance failures of quiescent components, but they may cause a deterioration of component strength under prolonged impact. If the quiescent component is in an operating system, in addition to the above-mentioned environmental stresses, it may also be exposed to mechanical stresses such as vibration, shock, and acceleration. These stresses are still environmental stresses which act on the component whether it is in operation or not, but they differ from the other stresses in that they can cause a chance failure of the component, causing also gradual fatigue and thus, deterioration of component mechanical strength.

---

* To gain assurance about the failure rates of components of a given make, the manufacturer has to supply the failure rate derating curves and the test results on which such curves are based.

2. Operating stresses which appear only when the component is in active operation such as voltage, frequency, current, self-generated heat, and, if the component or parts of it have to perform mechanical motions, as in relays, actuators, or rotating devices, self-generated mechanical stresses such as friction and vibration are also added. These stresses normally lead to chance failures, but at the same time they also contribute significantly to component strength deterioration which shows up in wearout.

From the point of view of an exponential reliability of electronic components in their useful life period, we are primarily interested in those stresses which can cause chance failures. These are mainly voltage and current transients, and high operating temperatures. Also included are shock, friction, and vibration, where they can occur above the component mechanical design strength or above the strength of the mounting arrangements. For instance, resonant frequency vibrations are to be avoided with utmost care. Components with their mounting arrangements have mass and elasticity and therefore they usually have several natural frequencies. These should be kept considerably higher than the highest forcing vibration frequency which may act externally on the component or its parts and mounting.

The chance effects of shock, friction, and vibration can be minimized or even completely eliminated by good mechanical design, mounting, and packaging or encasement. One very useful method is the potting of components in potting compounds which are good electrical insulators where necessary, and good thermal conductors. Good encasement or potting will also minimize the aging effect of the environmental stresses mentioned under (1), although it is known that complete protection against the penetration of moisture, corrosion, or radiation does not exist. The environmental stresses and their aging effects must be considered in connection with the mean wearout life and the shelf life of components. Components stored for a long time will absorb humidity, although in some cases this moisture can be "baked out" by drying; in other cases, prolonged humidity causes permanent damage. It is often said that some components, such as semiconductors, are more reliable after they have operated for thousands of hours than when they were taken from a storage shelf. In operation, temperature is often a good protection against humidity.

When investigating the effect of stresses on the chance failure rates of components, the first step is to determine those stresses to which the component failure rate is most sensitive. For electrical and electronic components, we find as a rule that temperature and electric stresses are primarily responsible for chance failures; for hydraulic components, line pressure; and as a rule for other components it is always that physical phenomenon on which the operation of the component is based. Temperature, also, must always be considered.

This discussion should enable us to derive more meaningful and realistic "failure rate vs. stress" curves for various kinds of components.

It has become a widely used practice in reliability analysis to derive electronic component failure rates from sets of curves similar to that in Figure 15.3 with a temperature scale only on the abscissa. The ordinate has a logarithmic failure rate scale usually in units of per cent failure per thousand hours of operation. Thus, a failure rate of 1 per cent per thousand hours corresponds to a $\lambda = 0.00001$ per hour or a mean time between failures of 100,000 hours, and a failure rate of 0.01 per cent per 1000 hours corresponds to $\lambda = 0.0000001$ per hour or a mean time between failures of 10,000,000 hours, etc. Each curve in such a graph represents, then, the component failure rate as a function of temperature at a given fixed voltage or wattage operating level.

It is important to realize that any sets of failure rate vs. voltage vs. temperature or failure rate vs. wattage vs. temperature curves apply at best to components of a definite type and manufacture operated at a given level of mechanical stresses. Another brand of the same type of components will display a different set of curves, and so will different types of the same manufacture. But the general pattern of the curves will be similar.

The dependence of component failure rates on temperature and dielectric stresses is often assumed to follow the *Arrhenius law* and the so-called *fifth power law*. The Arrhenius law states that the rate of chemical reactions of solutions is approximately doubled with each 10°C rise of temperature. From this the very rough rule was conceived that component failure rates are doubled with each 10°C temperature rise. The fifth power law is also only of a very rough rule-of-thumb nature: it states that the life of certain types of capacitors is inversely proportional to the fifth power of the operating voltage. This was construed to mean that component failure rates are directly proportional to the fifth power of the operating voltage.

A more realistic approach is to assume an nth power law for voltage changes, and a rule that failure rates are doubled with each T degrees of temperature rise. This approach yields the following formula for failure rate changes:

$$\lambda_2 = \lambda_1 \left(\frac{V_2}{V_1}\right)^n K^{(t_2 - t_1)} \tag{15.4}$$

where λ_1 is a known component failure rate at an operating voltage of V_1 volts and at a temperature of t_1°C, and λ_2 is the failure rate which the component assumes at an operating voltage V_2 and temperature t_2, the mechanical stresses being kept at the same level. The exponent n and the value of K must be determined experimentally and will change for each brand and type of components. K is a factor by which the failure rate changes with each 1°C temperature change. It is possible to assume that within the usual ratios of rated voltage to derated operating voltage and rated temperature to operating temperature, the operating values being always lower than the rated values, n and K can be regarded as constant for a given population of components. According to the kind and type of components, the values of n usually lie between 4 and 10, and the values of K between 1.02 and 1.15.

The experimental determination of n and K is done by first keeping the voltage constant at its rated value V_r and running a failure rate test at room temperature t_1 and then another test at the rated temperature t_2. This allows us to determine K from

$$\ln K = \frac{\ln \lambda_2 - \ln \lambda_1}{t_2 - t_1} \qquad (15.5)$$

Then a third failure rate test is run at room temperature t_1 but with operating voltage reduced to half the rated value, $V_r/2$, which yields a failure rate λ_3. From this test the value of n is determined as

$$n = \frac{\ln \lambda_1 - \ln \lambda_3}{\ln 2} \qquad (15.6)$$

With K and n known, the whole family of failure rate vs. voltage vs. temperature curves can be plotted, and from these curves the failure rates for a given operating voltage possible, from 0.1 up to 1.0 of the rated voltage and for a given operating temperature within the rated temperature limits, can be directly read. To supply these curves is a matter for the component manufacturer, who can determine the parameters K and n with three failure rate tests and plot the curves.

The same principles also apply when wattage is substituted for voltage, as is customary with components such as resistors, diodes, etc. Of course, caution is still necessary when such curves derived in the suggested way are then applied for failure rate determination, for two reasons: first, n and K are not absolutely constant over the whole range of temperatures and voltages. For instance, at operation above the rated voltage and above the maximum rated temperature, these coefficients can increase rather rapidly. But the assumption of their being constant is permissible for operation below the maximum rated values as well as above the rated minimum values. Second, strictly speaking, a set of such curves applies to a given level of the mechanical and other stresses. Thus, when such stresses are increased, the chance failure rates may also increase if affected by such stresses. However, hermetic sealing, potting techniques, and good mounting and packaging practices can greatly reduce these effects on the chance failure rates, so that when reliable mechanical design practices are used for assembling and packaging components in an equipment, it is permissible to assume that we are finally left with the electrical and thermal stresses as the main factors determining the failure rates of electrical and electronic components. As we said before, in hydraulic components the hydraulic stresses take the place of the electrical stresses, and in other components they are those stresses directly connected with the physical phenomenon on which the operation of the components is based. In such cases the operating to rated voltage or wattage ratio is replaced by operating to rated pressure, or other parameters according to the nature of the components, and again K and n can be determined and sets of curves plotted; however, modifications may have to be made in the above formulas.

Changing stress levels also affect the mean wearout life of components and its standard deviation—the effect is similar to that on the chance failure rates. Reduced stresses increase the mean wearout life but usually also increase the standard deviation of lives. Increased stresses reduce the mean wearout life and usually also reduce the standard deviation. This means that in preventive maintenance, components operating at higher stress levels must be preventively replaced earlier than the same components operating at lower stress levels.

Finally, in order to use failure rate vs. stress curves to the best advantage, the designer or the reliability analyst must have a good knowledge of the expected stress levels at which components will operate in an equipment or system. Thus a good stress analysis which predicts the operating temperatures of the individual components under various operating conditions and the expected magnitudes of electrical or other parameters, including transients, is needed. A thorough comprehension and study of the stress spectrum to which the individual components will be exposed will contribute greatly to the accuracy of every reliability analysis.

Chapter 16

SAFETY AND RELIABILITY OF
AIRCRAFT AND AIRBORNE SYSTEMS

COMPLEX SYSTEMS perform a multitude of functions which are necessary to assure the success of system operation. Each of these functions is performed with some degree of reliability, and the product of these reliabilities is the reliability of the system.

In designing a complex system for reliability, a quantitative reliability goal for the system is first established. This goal may be spelled out in customer specifications if the system is built for a specific customer. If the system is to be mass produced and offered to many customers, such as a computer, an automatic control system, or a passenger airplane, the manufacturer sets his own reliability requirements along with the performance requirements so as to build a marketable product which becomes a success in its use (does not suffer breakdowns in operation) and thus opens wider marketing opportunities for future products by establishing the reliability reputation of the manufacturer. In specifying reliability requirements, the cost of building to a reliability goal is weighed against the cost and effects of in-service breakdowns of the system. With a given reliability designed and built into a system, the average number of in-service breakdowns is also built into the system. It is very important to recognize this.

There is an additional factor which may increase the average number of system breakdowns in an unpredictable way—the human factor of the operating and maintenance personnel. This consideration, however, does not enter into system reliability design. On the other hand, the designer can and ought to design so as to minimize the possibilities of inadvertent operation and to make correct maintenance of the system as easy as

155

possible. In other words, he can make it difficult to operate the system inadvertently or to maintain it incorrectly. Why is this not a system reliability design consideration? Because reliability is defined as the probability of system operation without breakdowns caused by malfunctioning or failing system components, and this definition enables us to calculate system reliabilities in a realistic way. If human factors were included in the probability of successful system operation, a formula for system operational reliability would then result in the following form:

system operational reliability
  = system reliability
      × probability that operating personnel will not inadvertently operate the system
      × probability that the maintenance personnel has not made errors during the last maintenance operation

It is evident that these two human factor probabilities are above all a matter of personnel training, skill, intelligence, and goodwill and they belong in the category of pedagogy, sociology, and psychology rather than in technical engineering. Thus, when we speak of system reliability we have in mind the system and its components, and we minimize the probability of failure of these—although, as we said before, the designer can make it difficult for human errors to occur in system operation and maintenance. The designer should be like a good driver, who always anticipates that another driver or a pedestrian will make an error.

Quite apart from these human factors, the fact remains that the reliability for which a system is designed and built determines the system's functional mean time between failures and therefore the average number of system functional breakdowns. For a computer, for instance, this means that the average number of in-service failures is already built in, and for an aircraft it means even more. Not only is the average number of interrupted or uncompleted flights or emergency landings built in, but even the average number of accidents which will occur in a fleet of airplanes of that type and manufacture is predetermined by the reliability built into the airplane system.

In reliability considerations and in the establishment of reliability goals, there is a difference between systems which are subject only to an interruption of service if they fail and can be operated again as soon as the failed part or parts are replaced and systems where a failure or a combination of failures may cause the total loss of the system. The loss of the system is not only costly, but it may involve loss of lives, as in airplane accidents. Thus, in airplane design we are confronted with both the need to design for a certain reliability of completing flight missions and the need to design for safety. And safety, like reliability, can also be defined in probability terms. It is the probability that no accident will occur in a given length of flight time.

The same principles of probability calculations apply to both safety and reliability. Safety is simply the system's reliability that no accident caused by the system's failing or malfunctioning components will occur. The above definition enables us to treat safety as a reliability level at which certain functions must be performed.

From the reliability standpoint there are three types of equipment functions in an airplane:

1. Safety functions—performed by equipment which must operate on at least a specified minimum level to prevent aircraft accidents.
2. Essential functions—performed by equipment designed to ensure successful aircraft operation for a number of flight hours.
3. Convenience functions—performed by equipment intended to facilitate aircraft operation for the crew and to increase the comfort of passengers.

The reliability requirements for these three categories will differ considerably in magnitude. Category 1 includes the safety functions performed by the power plant (engines, including the fuel supply system), the controls (including brakes), landing gear, emergency electric power supply, certain electronic equipment, and certain navigational instruments. These functions must be performed with an extremely high reliability which is governed by crew-and-passenger-safety requirements rather than by system cost considerations.

Various figures have been suggested for mean flight hours between accidents as safety design goals, ranging from 100,000 to 1,000,000 flight hours and more, when all causes of failure which can result in an accident are considered. If we assume 1,000,000 hours as an absolute minimum safety design goal, this corresponds to an accident failure rate of

$$\lambda_{\text{accident}} = 0.000001$$

and to a minimum reliability of safety for 10 flight hours of the airplane, which must comprise the reliabilities of all safety functions, of

$$R_{\text{safety}} = 0.99999$$

This figure is the product of the reliabilities of all the equipment functions which must be performed on at least a minimum level to prevent an accident.

With this reliability built into an airplane, and assuming that perfect maintenance practices are employed and that the operating personnel is absolutely reliable, on the average one would expect less than one functional accident in 100,000 flights of 10 hours' duration each. In terms of a fleet of 100 of these passenger airplanes each operating 3500 hours per year and thus accumulating about one million flight hours in three years, the operator would expect not more than one accident in three years in his fleet. Because this amounts to an accident chance of 1:100,000 in one 10-hour flight and built-in safety of $R = 0.99999$, this also indirectly determines the insurance policies for that type of aircraft.

The safety design goal of the aircraft can be factorized in the following way into a product of the reliabilities of the safety functions:

$$R_{\text{safety}} = R(\text{aircraft structure}) \times R(\text{emergency propulsion})$$
$$\times R(\text{emergency controls}) \times R(\text{landing gear})$$
$$\times R(\text{emergency electric power})$$
$$\times R(\text{emergency electronics})$$
$$\times R(\text{emergency navigation instruments})$$

Because the structure failure rate* of well-designed structures can be assumed to be less than $\lambda_{\text{structure}} = 0.0000001$ per hour, the design goal of $R = 0.99999$ would apply to the product reliability of the propulsion, controls, landing gear, electronics, and navigational systems operating on at least such minimum level as to prevent an accident. With an equal reliability apportionment to these safety functions, it appears that each of them should be performed with a reliability of at least 0.9999984 for 10 hours of flight, or with correspondingly higher reliabilities for flights of shorter duration.

If we consider the engines from the viewpoint of the emergency propulsion requirement $R = 0.9999984$, we find that the ground rules for emergency propulsion change with the type of aircraft and its design. For instance, a twin-engine aircraft may be designed for safe landing on one engine, whereas most four-engine aircraft require two engines for safe landing. If we assume a jet engine failure rate of $\lambda = 0.0001$ per hour, or with the fuel supply system included, $\lambda = 0.00011$ per hour, the expected emergency propulsion reliability of a two-engine system amounts to 0.9999988 for 10 flight hours if at least one out of two engines must remain operative. This results from simple parallel redundancy considerations. For a four-engine system, in which at least two out of four engines must remain operative, the emergency propulsion reliability amounts to $R = 0.9999999$ for 10 flight hours. Thus, from the point of view of the specified requirement for the propulsion reliability for safe landing of $R = 0.9999984$ for 10 flight hours, both systems appear to be satisfactory.

Now the landing gear, essential controls, emergency electric power supply, essential electronics, and essential navigation instruments also require a reliability of the order of magnitude of $R = 0.9999984$ each.

---

* John J. Coleman, "Reliability of Aircraft Structures in Resisting Chance Failures," *Operations Research*. 7:5, 639 (Sept.–Oct. 1959).

As in the propulsion system, these high reliabilities can be achieved only by the use of parallel or stand-by redundant systems.

However, caution is necessary in the calculations where one system depends on the operation of the other system, as, for instance, in the case of the emergency electric power supply. The function of supplying emergency electric power may require that at least one generator remain in operation. With one generator on each engine, the probability of engine failure must also be considered, because we are not concerned here with the probability that at least one out of two or four generators will not fail, but with the probability that electric power will be available from at least one generator. These two probabilities are not identical, because in specifying the function of "electrical power being available," we at the same time require that at least one engine remain operative and that the generator mounted on the surviving engine not fail. In other words, in calculating the probability that electric output from at least one generator will be available, we enter the engine-generator sets into our parallel reliability calculations, and not only the generators. Obviously, we would get the wrong answer if we considered only the generators, as can be seen immediately from the twin-engine aircraft example: If generator b mounted on engine B fails, and engine A also fails, we have no electric power left even though generator a may still be in perfect condition.

Let us consider the partial reliability

$$R'_{\text{safety}} = R(\text{emergency propulsion}) \times R(\text{emergency electric power})$$

which pertains to propulsion and electric power only. For the twin-engine aircraft we have $R(\text{emergency propulsion}) = 0.9999988$, and we require that the emergency electric power supply function be

$$R(\text{emergency electric power}) = 0.9999984$$

for electric power being available from at least one generator. The block diagram in Figure 16.1 facilitates the calculation.

Assuming a failure rate of $\lambda = 0.00049$ for a generator, including its regulating equipment and drive, the reliability that electric power will be available from at least one generator in a 10-hour flight is

$$R = 1 - (1 - R_e R_g)^2 = 0.999964$$

But the engines are included in this reliability evaluation, and therefore the obtained result equals our partial reliability R'_{safety} and amounts to the probability that both emergency propulsion and emergency electric power will be available. Therefore, we can write

$$R'_{\text{safety}} = R(\text{emergency propulsion}) \times R(\text{emergency electric power})$$
$$= 0.999964$$

But now what is the reliability of the emergency electric power function alone, or what is the actual value of R(emergency electric power) which we require to be at least 0.9999984 according to the equation for aircraft safety R_{safety}. This amounts to the conditional reliability that emergency

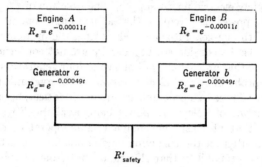

Fig. 16.1. A safety block diagram.

electric power will be available given that at least one engine survives. The probability that one engine out of two survives is

$$R\text{(emergency propulsion)} = 1 - (1 - R_e)^2 = 0.9999988$$

and therefore

$$R\text{(emergency electric power)} = \frac{1 - (1 - R_e R_g)^2}{1 - (1 - R_e)^2} = 0.999966$$

It is this value which by the reliability apportionment to the emergency electric power safety function is required to be 0.9999984. As we see, this figure is with the given generator failure rate not achieved and therefore either a battery of sufficient rating would have to be used if the electric system were d-c, or if a-c, one could put two generators on each engine.

In a four-engine configuration, where for safe landing at least two engines and a generator on one of the surviving engines must operate, we obtain, for the given failure rates

$$R'_{\text{safety}} = R_e^4(1 - Q_g^4) + 4R_e^3 Q_e(1 - Q_g^3) + 6R_e^2 Q_e^2(1 - Q_g^2) = 0.9999998$$

The reliability of emergency propulsion for at least two out of four engines operating is

$$R_{\text{(emergency propulsion)}} = R_e^4 + 4R_e^3 Q_e + 6R_e^2 Q_e^2 = 0.9999999$$

Therefore the reliability of emergency electric power is

$$R_{\text{(eep)}} = \frac{R_e^4(1 - Q_g^4) + 4R_e^3 Q_e(1 - Q_g^3) + 6R_e^2 Q_e^2(1 - Q_g^2)}{R_e^4 + 4R_e^3 Q_e + 6R_e^2 Q^2} = 0.9999999$$

which is better than the required 0.9999984. To design the generating system exactly to this required goal of 0.9999984, we could thus relax on the reliability of the generators when the engine failure rate is $\lambda = 0.00011$. The generator minimum required reliability R_g can then be calculated from the above equation by equating it to the required value of 0.9999984 of the emergency electric power safety, and obtaining from R_g the minimum required generator mean time between failures or its reciprocal, the maximum allowable generator failure rate.

Wherever dependence of functions is involved, conditional probabilities must be considered in the evaluation of the reliabilities of the individual safety functions. A similar procedure as outlined for the emergency propulsion and for the emergency electric power supply is then used in determining the required reliabilities of the equipment providing for the safe operation of the landing gear, emergency controls, emergency electronics, and emergency navigation. The calculations for apportioning reliabilities to the individual equipments when R_{safety} is specified for the whole airplane, or for calculating in a reliability analysis R_{safety} from the reliabilities of the various equipments and their components, follow fundamentally the same principles and procedures outlined above.

Safety considerations and calculations can also apply to unmanned vehicles—missiles, for instance—to minimize their chance of hitting friendly territory by minimizing the probability of certain malfunctions or failures.

Next, we consider category 2, i.e., the essential functions which must be performed in an aircraft to complete a flight or mission successfully. These are the functions which determine aircraft reliability in the conventional sense. If any of these functions fail, a decision not to continue the flight would be made, which involves emergency landing. The reliabilities of these functions must also be high because, apart from the cost of such emergency landings, the reputations of the aircraft operator and manufacturer are involved, which are not readily measured in terms of money.

These reliabilities are governed by the maximum permissible rate of flight abort occurrences. The reliability requirement may vary from operator to operator.

There are certain basic requirements which, though dependent in part on the type of aircraft, are usually those listed below.

(a) Essential propulsion power (at least two out of two or three out of four engines must operate to justify the continuation of a flight to its final destination), must be generated.

(b) Essential electric power (at least x out of y generators must provide electric power, and the essential loads must be supplied and operative), must be generated.

   (c) All controls, including brakes must operate with redundancy
       built into their operation.
   (d) Essential communication equipment must operate (at least one set
       of these must operate; redundant equipment may fail).
   (e) Essential navigation instruments must operate (redundant instru-
       ments may fail).
   (f) Pressurization must be maintained at a tolerable level.
   (g) Temperature must be maintained within tolerable limits.
   (h) Fire detection system must operate.
   (i) Fire extinguishing system must operate.
   (j) Hydraulic system must operate.

These functions must be performed at the indicated essential levels in
order to complete a flight successfully. The reliability of the landing gear
has been considered already under safety. Having set these ground rules,
we can now define the reliability of an aircraft as the probability that
all essential functions will be performed at the required levels from the
time of departure from one airport to the time of arrival at the next
scheduled airport. This probability is equal to the product of the probabilities with which the individual essential functions will be performed
for a required flight time t:

$$R_{\text{aircraft}} = R_a R_b R_c \cdots R_j$$

If any of these functions (a) through (j) cannot be performed at the required levels, an emergency condition arises and a landing at the nearest
airport is required. Some of the function reliabilities will be of a conditional probability nature—for instance, the supply of essential electric
power which depends on the engines, or the functions of some other subsystems which may depend on the operation of the electric generating
system, etc. In such cases, conditional reliabilities may have to be considered in the reliability apportionment of aircraft reliability analysis,
similar to the case of aircraft safety evaluation.

   The reliability design goal of an aircraft will depend on the acceptable
flight abort rate, which may be of the order of magnitude

$$\lambda_{\text{abort}} = 0.001 \text{ to } 0.0001$$

Assuming that the abort rate were specified not to exceed 0.001, i.e.,
not more than one aborted flight per 1000 flight hours (not an exceed-
ingly high aircraft reliability requirement), this would require the aircraft
to be designed to a reliability of at least 0.99 for flights of 10 hours'
duration (or 0.995 for 5-hour flights, etc.). And because there are 10
essential functions, and if equal apportionments are made to these (equal
apportionments are used only for simplifying the example), each would

have to be performed with a reliability of 0.999 for 10 hours (or 0.9995 for 5 hours) so that $0.999^{10} = 0.99$ (or $0.9995^{10} = 0.995$, according to the average length of flight for which the aircraft is being designed).

For instance, the reliability of the essential propulsion power for 10 hours of a twin-engine jet aircraft would be only 0.9978 if the engine hourly failure rate is 0.00011 and both engines must operate to continue a flight. On the other hand, the reliability of a four-engine jet would be 0.99999 if it is specified that the flight can continue to its destination with only three out of four engines operating, i.e., with one failed engine. (Modern jet engines have lower failure rates than the failure rate used in this example.)

The above reliability considerations apply to nonstop flights between terminals, where sufficient time is available for maintenance. The situation is different when flight schedules include intermediate stops. For instance, if one out of four engines is lost on a direct flight, the aircraft could continue flight to reach the scheduled airport, without undue delay. But if an intermediate stop is scheduled and the failure occurs before this intermediate stop is reached, the airplane will fly in but will not be allowed to take off with only three engines operative.

This would cause a delay, or an abort of that number flight. On the other hand, if the engine fails on the second lap of the flight, i.e., between the intermediate stop and the terminal, then no delay need occur. In terms of aircraft propulsion system reliability, this means that up to the last intermediate stop all four engines must operate and at least three out of four engines must operate on the last lap.

In general, if a flight is not allowed to continue from an intermediate stop unless the essential functions of the airplane are operative with all the redundancy built into them (such as the requirement that all four engines must operate), then the aircraft's probability of successfully completing a flight between two terminals in time t depends on the component failure rate, i.e., the series failure rate, of the equipment performing these specific functions in the flight time t_1 up the last intermediate stop, and on the flight abort rate alone for the last lap in flight time t_2,

$$R = [\exp{(-\lambda_{components}t_1)}] \cdot [\exp{(-\lambda_{abort}t_2)}]$$

where $t_1 + t_2 = t$. This means that the essential functions must then be performed with a higher order of reliability (no failure of redundant elements allowed) in the time t_1 than in the time t_2, if in addition to preventing flight aborts the requirement is added that no delays should occur at the intermediate stops. It seems that in practice the strict observance of these rules is limited to engine failures.

Finally, there is category 3. These are convenience functions which, if lost, will not preclude a flight from normal continuation. Their loss does not necessitate a decision for emergency landing. These functions are performed by certain auxiliary equipment, and may be auto-pilots, the water injection system, some refined functions of air conditioning and exact temperature regulation, and most convenience items. The reliability of such functions is not connected with aircraft safety or reliability and is governed by such considerations as the operation and maintenance cost. The reliability level built into them depends on the importance assigned to them by the operator. From a maintenance point of view, the simpler these systems are, the less maintenance they will require.

In the following chapter we shall deal with maintenance considerations of aircraft systems and with system maintainability, which is of general importance for aircraft utilization and availability.

Our discussion on system safety and reliability reveals an important fact. The worth of systems which, because of their operational features, require that due consideration be given to safety, is defined jointly by the probability of safe operation and the probability of mission completion. Thus, to define an aircraft's airworthiness two figures must be specified—the aircraft's reliability $R_{aircraft}$ and its safety R_{safety}. These jointly determine the reliability which must be designed into the subsystems and equipments performing the individual functions.

In addition to the airworthiness of an aircraft, its ability to give continuous service in accordance with predetermined operational time schedules must also be considered as a design parameter. Thus, service ability, also called *servicability*, determines the worth of a system in its use for prolonged periods which can be expressed in terms of a system utilization factor. System maintainability plays an important role in this connection.

Chapter 17

SYSTEM MAINTENANCE,
AVAILABILITY, AND DEPENDABILITY

ALL RECOVERABLE SYSTEMS which are used for continuous or inter-mittent service for some period of time are subject, at one time or another, to maintenance.

Maintenance actions can be classified in two categories: First, there is *off-schedule* maintenance, necessitated by system in-service failure or mal-function. Its purpose is to restore system operation as soon as possible by replacing, repairing, or adjusting the component or components which cause interruption of service. Second, there is *scheduled* maintenance at regular intervals; its purpose is to keep the system in a condition con-sistent with its built-in levels of performance, reliability, and, where applicable, safety.

Scheduled maintenance fulfills this purpose by servicing, inspections, and minor or major overhauls during which

(1) regular care is provided to normally operating subsystems and components which require such attention (lubrication, refueling, cleaning, adjustment, alignment, etc.),
(2) failed redundant components are checked, replaced, or repaired if the system contains redundancy, and
(3) components which are nearing a wearout condition are replaced or overhauled.

These actions are performed to prevent component and system failure rates from increasing over and above the design levels. Therefore, sched-uled maintenance is also called *preventive* maintenance.

The frequency at which maintenance actions of type (1) must be per-formed to prevent system reliability degradation depends on the physical characteristics of the components involved; for type (3) items it depends on the wearout statistics of the components and their number in a system, as explained in Chapters 6, 7, and 8.

Therefore, the frequency of type (3) maintenance actions will differ for various types of components, but optimum preventive maintenance timetables can be established for every system in advance; the manhours and the length of time required by these maintenance actions can also be determined.

The frequency of type (2) maintenance actions is of a different character because it is governed by probabilities. It depends on the failure rate of the components in a redundant system and on the reliability requirements at which such a system must operate. This dependency will be explained later. For the time being, let it be accepted that a fixed timetable can also be scheduled for this type of maintenance. The timetable consists of prescribed inspection periods for the purpose of finding out whether there are failed components in a redundant system, because failures of redundant components do not necessarily show up if special indicating provisions are not incorporated. If failed components are detected, they must be replaced or repaired to restore the system to its design reliability level. The time required for such inspection and maintenance actions can be estimated in manhours and in clock hours and allocated in advance as an additional time to be added to the fixed times provided for types (1) and (3). Alternately, it can be treated as a discrete variable independent of the fixed time schedules. For simplicity we prefer the first approach here so that a fixed total clock time T_p is scheduled for all preventive maintenance actions together, for every t system operating hours.

As to the frequency of the off-schedule maintenance, this is strictly a function of the failure rates of those components or units which cause in-service failures of the system, and therefore it is a function of the reciprocal of the system's mean time between failures m.

For every t system operating hours there will be on the average t/m in-service system failures, and therefore on the average t/m off-schedule maintenance actions will have to be performed. Off-schedule maintenance is also called *corrective* maintenance. The manhours required for these repairs will vary with the components which cause the failures. The t/m off-schedule maintenance actions can be broken down into t/m_1, t/m_2, t/m_3, etc., i.e., the number of maintenance actions to be performed on the individual components and units which have mean times between failures m_1, m_2, m_3, etc.* The total average number of off-schedule

---

* All mean times between failures must be expressed in terms of the system operating time as explained in Chapter 10, and those of regularly maintained redundant units are derived from Equation (20.7) in Chapter 20.

maintenance actions for t system operating hours is thus given by

$$\frac{t}{m} = \frac{t}{m_1} + \frac{t}{m_2} + \frac{t}{m_3} + \cdots = \sum_{i=1}^{n} \frac{t}{m_i} = \sum_{i=1}^{n} \lambda_i t \qquad (17.1)$$

Now if the time in manhours required for repairing a system failure caused by component 1 (which has a mean time between failures of m_1) is known to be T_1 including so-called secondary failures which may be present, and for component 2 it is T_2, etc., then the total average manhours H_o involved in off-schedule maintenance for every t system operating hours become

$$H_o = \frac{t}{m_1} T_1 + \frac{t}{m_2} T_2 + \frac{t}{m_3} T_3 + \cdots = \sum_{i=1}^{n} \frac{T_i}{m_i} t = \sum_{i=1}^{n} (\lambda_i T_i) t \qquad (17.2)$$

The individual times T_i which represent the times needed to restore system operation in each particular case depend on the physical location and design characteristics of the components in the system, including their accessibility, their mountings, and the provisions made to detect the component which has failed. The times T_i may become longer if unskilled personnel is used, in which case each T_i has to be multiplied with a factor larger than unity. This factor can be defined as the *reciprocal of personnel efficiency or skill*.

Assuming that H_o in Equation (17.2) is the total average time in manhours required for a system's off-schedule maintenance with personnel of a given skill, and that it includes fault location before repair and checkout after repair, it can be converted into a total clock time T_o according to the available manpower. When this time T_o is added to the total time T_p allocated to system preventive maintenance, we obtain the total maintenance clock time T_m which has to be spent on the average for each t system operating hours:

$$T_m = T_p + T_o \qquad (17.3)$$

This also represents the system's functional down time. Thus, with a provision made for down time other than functional designated as T_r, which may be the scheduled idle periods for t system operating hours or other scheduled activities such as administrative time, etc., we obtain the system utilization factor:

$$U = \frac{t}{T_p + T_o + T_r + t} \qquad (17.4)$$

The numerator is the time the system will be expected to operate for the calendar time, $T_p + T_o + T_r + t$, which is in the denominator. It is customary to choose one year or 8760 hours as calendar time.

The important factor which determines maximum possible system utilization is the total maintenance time or functional down time $T_m = T_p + T_o$ for t system operating hours. Therefore, the maximum possible system utilization factor is

$$U_m = \frac{t}{T_m + t} \tag{17.5}$$

for a system which requires on the average T_m hours of maintenance for t operating hours. To increase this factor necessitates a reduction in T_m at the design stage. This can be done in several ways—by increasing the system's mean time between failures, which will reduce the number of in-service failures given by Equation (17.1) and, in addition, by reducing the times T_i in Equation (17.2). These measures will result in a reduction of the manhours H_o and the time T_o required for off-schedule maintenance. The most effective design procedure is to reduce the times T_i of those components which have the lowest mean times between failures m_i, which means to locate these components in the system so that they will be easily accessible and to see that their mounting arrangements will allow fast replacement. Also, of course, if possible one should obtain components with lower failure rates or provide for derated operation because every reduction in component failure rates as well as in component replacement times T_i will reduce the over-all off-schedule maintenance time T_o of the system.

As to the scheduled regular maintenance time T_p, this can be reduced by the use of self-lubricating, hermetically sealed components which do not drift out of tolerance while in use, and which have long mean wear-out lives with as little deviation from the mean as possible. Here again, components which require regular attendance or frequent replacement because of a high wearout rate should be easily accessible—the more maintenance they require, the easier it should be to get to them. Spectacular increases in the system utilization factor can be achieved when proper consideration is given to maintenance in the system design stage.

Where parallel redundant components are used, the sum of their failure rates determines the number of maintenance actions required. If we consider two components in parallel each having the same failure rate λ, the parallel system mean time between failures will be $3/2\lambda$ hours. Thus, in-service system failures would occur on the average once every $3/2\lambda$ system operating hours if no replacement is made when one of the two components fails, i.e., if the system is allowed to operate until the surviving component also fails.*

---

* See Chapter 20 for the calculation of the mean time between failures of regularly checked and maintained redundant systems.

On the other hand, the mean time between failures of one component is $1/\lambda$, and with two of these components in parallel there will be on the average a component failure in the parallel system once every $1/2\lambda$ system operating hours if the failed components are replaced immediately so that two components are always operating. With a policy of immediate replacement of failed redundant components, redundant systems can be built whose mean time between failures will approach infinity.

This brings us to the space-technological aspect of system reliability and maintainability design. If a redundant system is so designed that the surviving component can continue to operate while the failed component is being replaced—if the system operation does not need to be interrupted during the replacement action—system reliability can be made entirely independent of the system operating time t. Assume that the act of replacing a failed redundant component requires a short time τ. With the other component operating during this replacement action, the parallel system could fail only if during the short time τ the operating component also failed. The probability of this happening is $1 - e^{-\lambda\tau}$, and when τ becomes infinitely small, i.e., when theoretically immediate replacement takes place, we get $1 - e^{-\lambda 0} = 0$, which means that the system would never fail.

Although it is not realistic to assume replacement in time $\tau = 0$, systems can be designed so that they continue to operate while failed redundant components are being replaced, and τ can be made very short by plug-in or similar arrangements and by the use of fail-safe indicating circuitry. The reliability of such a two-component parallel system then amounts to the probability that the operating component will not fail in the time τ measured from the failure of the other parallel component to its replacement:

$$R(\tau) = e^{-\lambda\tau} \tag{17.6}$$

Thus, the system reliability becomes virtually independent of the system operating time t and depends only on the short replacement time τ.

If three components are operating in parallel in such a redundant system so that one surviving component can continue the operation, the reliability of the system becomes

$$R(\tau) = 2e^{-\lambda\tau} - e^{-2\lambda\tau} \tag{17.7}$$

which is again independent of the system operating time. Thus, for all practical purposes such systems can be considered as absolutely reliable when τ is short.

The use of these systems with a reliability independent of the system operating time is warranted where continuity of operation is highly essential. This is one way to solve the reliability problem of absolutely essential equipment in manned space vehicles for interplanetary and space flights of many years' duration, or for computers and other vital electronic devices whose continuous, nonstop operation is an absolute must.

Obviously, a system whose reliability is made independent of the operating time requires in-service attendance and maintenance. The problem here is that of carrying the required number of replacement spare parts and/or repair kits to allow for fast repair of failed components. The number of expected in-service maintenance actions depends on the operating time and on the reliability of the components used. Because components here are not replaced preventively but are allowed to operate until they fail, which amounts to their maximum possible utilization and includes wearout, the component mean time between failures amounts to

$$m = \int_0^\infty L(t) \, dt \qquad (17.8)$$

where $L(t)$ is given by Equation (8.22) in Chapter 8. In such situations wearout has to be considered in addition to any chance failure possibility. The knowledge of the component's m allows us to compute the expected number of maintenance actions for each component over a prolonged system operating period t, and with adequate consideration given to the probability that more than the expected number of component failures might occur,* it is then possible to compute the optimum number of spare parts or the kind of repair kit to be carried. In manned space vehicles this approach would be necessary for those functions whose nonstop operation is essential for crew survival.

When in-service maintenance is impossible or component failure indication is impractical, components in redundant parallel systems must be periodically inspected to be sure that none of them has failed, and that the system still has its parallel reliability. Such inspection is sometimes carried out after every operation of t hours duration, and in other cases after n operations, i.e., after $T = nt$ hours. Let us assume that the group of parallel components is part of a larger system which is required to have a reliability $R_s(t)$ for t hours, and consistent with this requirement, the parallel group is required to have a reliability $R_p(t)$ for the same t hours. We know that the reliability of a parallel group of equal components is $1 - [Q(t)]^n$, but only if at the beginning of the operating period of t hours all components are in a good condition. This requires inspection of the parallel group before each operation and replacement of parallel components which might have failed in the preceding operation.

---

* Guy Black and Frank Proschan, "Spare Part Kits at Minimum Cost," *Proceedings*, Fifth National Symposium on Reliability and Quality Control, p. 281 (January 12–14, 1959), describes the case of chance failures. Also see Chapter 22, last example.

If inspection or replacement does not take place, the next operating period t may start with one or more parallel components failed. This may go on unnoticed as long as the last of the parallel components has not failed. But each failure of a parallel component reduces the reliability of the group, and therefore the reliability of the system. Therefore, if the system is required to have a specified reliability of $R_s(t)$ for each subsequent operation, but inspection of the parallel group of components in that system is scheduled only for every $T = nt$ hours, i.e., after every n operations of the system, the parallel group must be designed for a reliability $R_p(T)$, because only then can the system be expected to have the required reliability $R_s(t)$ for any operation of t hours of duration. On the other hand, if inspection takes place after every operation, the parallel group can be designed for a reliability $R_p(t)$. This is shown in Figure 17.1.

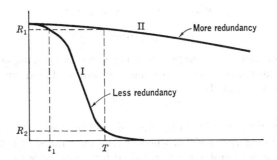

Fig. 17.1. Inspection and redundant system reliability.

As to the expected average number of maintenance actions, i.e., the average number of components which have to be replaced in a parallel group over some long period, for instance, in 1000 operations of length t each, we find this by expanding the binomial $(r + q)^n$ of the parallel group of n components where r and q stand for the reliability and unreliability of one component for the operating time between two inspections, which may be t or $T = nt$, according to the chosen inspection period. The procedure of computing the expected number of failed components has already been explained in Chapter 11. This procedure also allows us to compute the average number of inspections in which no maintenance will be required because none of the parallel components has failed, and the average number of inspections which result in finding one, two, three, etc., failed components. By adding these up we obtain the average number of components to be replaced in a parallel group. This, again, makes it possible to plan for a number of components to be carried as spares, in stock, considering the probability that in a given time more than the expected average number of failures may occur. It also allows us to estimate the amount of maintenance which will be required over a given time.

For example, consider a parallel group of three equal components which is inspected for component failures every T hours. Let the reliability of a component be r for T operating hours and its unreliability, q. Expanding the binomial of the parallel group, we obtain

$$(r + q)^3 = r^3 + 3r^2q + 3rq^2 + q^3$$

In a large number of inspections, say n inspections, we would then expect to find no failed component in r^3n inspections, to detect one failed component in $3r^2qn$ inspections, two failed components in $3rq^2n$ inspections, and q^3n times the whole group would have failed, i.e., all three components failed. The total average number N of failed components which have to be replaced in n inspections then amounts to

$$N = n(3r^2q + 2 \cdot 3rq^2 + 3q^3)$$
$$= n(3r^2q + 6rq^2 + 3q^3) \tag{17.9}$$

If the manhours to completely replace one failed component are h_1, the total manhours required for replacements of failed components in n inspections amount to

$$h_{\text{total}} = Nh_1$$

And if we add the manhours h_2 required to check the components during any one inspection to find out whether they have failed or not, so that in n inspections nh_2 manhours are required for checking, we obtain the expected over-all manhours required for the entire maintenance procedure in n inspections:

$$h_{\text{over-all}} = Nh_1 + nh_2 \tag{17.10}$$

The above calculation applies when component reliabilities do not change with component age. If replacement of unfailed components is necessary to prevent wearout, this falls into the category of regular scheduled overhauls, i.e., type (3).

According to the manpower which can be allocated to perform these maintenance procedures, the manhours $h_{\text{over-all}}$ can be converted into straight clock hours C. Thus, with n inspections scheduled for every t operating hours, the parallel group has an average repair rate of C/t hours per one system operating hour, and because $t = nT$, where T is the scheduled time between inspections, the repair rate becomes C/nT.

When several maintenance actions are being performed simultaneously on a system by a crew, and if all these maintenance actions start at the same time, the maintenance clock time required for restoring system to operation is determined by the longest maintenance action. In this case the shorter maintenance actions do not enter into the clock time,

SYSTEM MAINTENANCE, AVAILABILITY, DEPENDABILITY   173

or down time, computation at all. But the number of manhours required by each individual action is important because it is a measure of the total maintenance effort which has to be expended and allows us to plan for the optimum manpower and to establish optimum maintenance charts, so that the maintenance clock time, or down time, can be minimized.

To coordinate system maintenance time with system operating time, and therefore maintenance with reliability, the knowledge of the average functional system down time, i.e., of the average maintenance clock time, is necessary.

We have shown that the average maintenance manhours connected with in-service system failures and with component failures in redundant systems can be estimated from the reliabilities of the components, and that the average down time can be evaluated from maintenance charts. The various procedures followed in setting up such charts and optimum maintenance schedules are not the subject of this book. We are interested here in the interaction between reliability of a system and its maintenance.

In Equation (17.5) we defined the maximum possible system utilization factor, $U_m = t/(T_m + t)$. This factor is a measure of the system's availability because it gives the percentage of time the system will be available for operation. We have seen that to operate for t hours out of a total time of $T_m + t$ hours, the system requires T_m hours for scheduled and off-schedule maintenance. If for the system operating time we select its mean time between failures m, which may be some fraction or some multiple of t in U_m, we can then derive the average maintenance time T'_m which is required for every m system operating hours. Because in the evaluation of a system's utilization factor it is customary to choose one year for $T_m + t$, i.e., 8760 hours, so that T_m is the clock time in hours spent per year on system maintenance, the average maintenance time T'_m per m system operating hours will be

$$T'_m = T_m \frac{m}{t} = \frac{mT_m}{8760 - T_m} \qquad (17.11)$$

where m is the system's mean time between failures for a given maintenance policy, and T_m the system down time per year. When we now use m and T'_m instead of t and T_m in the utilization factor, we obtain a value which is numerically identical with U_m, which is by definition called system availability A:

$$A = \frac{m}{m + T'_m} \qquad (17.12)$$

It gives the same percentage of average time the system will be available for service as obtained from U_m, which we have defined as the maximum possible system utilization factor.

The highest value A can have is unity, or 100 per cent. This occurs only if a system never requires maintenance and therefore if T_m and T'_m are zero. When analyzing A we find that it is a ratio of system operating hours to the sum of system operating plus maintenance hours. By its nature, therefore, this ratio, called system availability A, is a probability as defined at the beginning of Chapter 4. System availability A is thus the probability that a system which has a mean time between failures m and requires a maintenance time T'_m for each m operating hours will be available for operation at any given time in the future. Conversely, we can also define a complementary probability B so that $A + B = 1$, and B is the probability that the system will not be available for operation. This probability, called *system unavailability*, amounts to

$$B = \frac{T'_m}{m + T'_m} \tag{17.13}$$

We have given Equations (17.12) and (17.13) because they are often used in literature dealing with maintenance problems, but both (17.12) and (17.13) can be expressed and also numerically evaluated in a much simpler way, by substituting for T'_m the system's functional down time T_m per year. We then obtain for system availability

$$A = 1 - \frac{T_m}{8760} \tag{17.14}$$

and for system unavailability

$$B = \frac{T_m}{8760} \tag{17.15}$$

This procedure eliminates both system operating time t and system mean time between failures m from the equations for system availability and unavailability. This does not mean, however, that A and B are independent of the system mean time between failures. It only means that the factor which is really affected by m is the system average maintenance time T_m, and its dependence on m was explained at the beginning of this chapter and is shown by Equations (17.1), (17.2), and (17.3).

Sometimes it is required to know the probability that a system will be available for operation between any two scheduled maintenances, i.e., when assigned to continuous duty so that it must be available at a moment's notice or is required to operate without interruptions. We call this probability *system dependability* D. It is derived in the same way as system availability in Equation (17.14), only we leave out from T_m, in Equation (17.3), the term T_p which stands for average scheduled maintenance time, and consider only the term T_o which stands for average off-schedule maintenance time, and take this on a per year basis. We then obtain for system dependability

$$D = 1 - \frac{T_o}{8760} \qquad (17.16)$$

Again, T_o can be evaluated from Equation (17.2) which gives the average off-schedule maintenance manhours H_o per t system operating hours in terms of component reliabilities, and this can be converted into T_o for a given crew and maintenance chart. Or, from (17.2) we first derive

$$\frac{H_o}{t} = \sum_{i=1}^{n} (\lambda_i T_i)$$

which is the number of off-schedule manhours required per 1 hour of system operation, and convert this into $r = T_o/t$ which represents the off-schedule clock hours, or proportionate repair down time, per 1-hour system operation. System dependability, then, can also be expressed as

$$D = \frac{1}{1 + r} \qquad (17.17)$$

Whichever form is used for system dependability, our main purpose is to perform a system maintenance analysis which will allow us to compute H_o for t system operating hours, or H_o/t per 1 system operating hour. Therefore, the failure rates λ_i of the components which can cause system failure must be known, and the times T_i required for restoring system operation when component i fails must be estimated according to the complexity of the component, its location and mounting in the system, and whether it can be replaced or has to be repaired. For the purpose of such analysis, redundant groups or subsystems are considered as units which have an over-all mean time between failures, the reciprocal of which is the whole unit's average failure rate.* It is important to note that the times T_i do not depend on component failure rates, but on system maintainability design and on the physical characteristics of the components such as their complexity, mounting, location in the system, etc.

* Given by Equation (20.9), Chapter 20.

Chapter 18

RELIABILITY
DESIGN CONSIDERATIONS

R ELIABILITY WAS DEFINED as "the probability of a device performing its purpose adequately for the period of time intended under the operating conditions encountered." Thus, adequate performance, which is a prime objective in every engineering design, is also one of the main postulates in the definition of reliability. Reliability requires adequate or satisfactory performance and also adds two more requirements—the performance must be maintained for a given period of time in a certain operating environment, and some probability, usually very close to 100 per cent, is specified that the equipment will not fail during the given time.

The designer is thus faced with two problems—to design the equipment for performance, which means to design it so that it will perform the purpose for which it is needed, and to design it for endurance so that it will not fail in operation.

Every designer knows that there are always several ways to design an equipment or a system for a specified performance. But the choice is narrowed down when reliability requirements must be met in addition to performance requirements. Usually one specific design will be more reliable than the other possible approaches.

The designer who has to meet specified reliability requirements should, throughout his design work, think in terms of several approaches to designing for performance. Naturally, performance of the designed equipment is the main objective—the objective which must be met first. If the performance requirements cannot be met, then there is no point in wasting time on any other considerations. But normally performance requirements can be met, and the designer should think of several ways to do it.

176

At this stage, before a definite approach is decided upon, a preliminary reliability analysis will give a good indication which of the possible approaches promises to have the highest inherent reliability, or in which of these approaches the reliability requirements can be easiest met. That is, then, the approach to final design which must be chosen. This design philosophy, although it requires more thought at the initial stage, has proven to be the most rewarding. It saves time and money, and may well be the deciding factor in contract negotiations and in whether or not the equipment will be a sales success.

In reliability design work, one should never be satisfied with just one idea of how to do a job. Your first idea may be the very one out of a number of possible approaches that is the least reliable, or the one that makes it most difficult to design the required reliability into the equipment. Be critical towards yourself and be selective. If you see one approach and continue to give more thought to the problem, you will soon see several approaches, and when selecting one of them, do not let yourself be guided by what is called "intuition" as to which approach is more reliable. Only a probability analysis can show this, and the approach which you intuited to be the most reliable may turn out to be the least reliable approach or the approach in which reliability is most difficult to achieve. You will do well to remember that complex probability (and reliability is a probability of a more complex nature) usually completely defies intuition.

Whether an equipment will meet a quantitatively specified reliability is essentially decided during its design stage. Quantitative reliability design, especially when complex equipment is to be built, requires a reliability control program effective during the entire design stage and extending beyond it to prototype approval and production.

The purpose of such a program in the design stage is to review design progress at predetermined phases and to ensure that reliability requirements are being met. Such a reliability control program has to be set up individually, from case to case, according to the complexity of the equipment involved. Essentially, however, it covers at least three phases of design work.

Phase 1 is the preliminary reliability analysis of tentative design alternatives resulting in the choice of the alternative which then goes into final design work.

When detailed design work is completed, phase 2 of the reliability program begins. It consists of a precise reliability analysis of the final design to assure that the reliability specifications have been met by the design.

Phase 3 of the reliability control program consists of testing the first prototypes and correlating test results to phase 2 analysis. If the correlation is good, no redesign is necessary and release for production follows. If correlation is not satisfactory, redesign may be required for the critical areas where failures tend to occur on the tested prototypes. The more thorough the analyses of phases 1 and 2, the less likely it is that redesign will be required.

The preliminary reliability analysis of phase 1 applies to proposals as well as developmental contracts. It is a theoretical comparative analysis of the various possible approaches and various alternatives tentatively proposed by the designer. Circuit and stress analyses are made which help to select component failure rates for the reliability analysis. Then reliability block diagrams are made for the alternatives to show where components combine in series and where redundant elements occur, and a reliability calculation follows to evaluate the $R(t)$ of the equipment for the specified mission time t, or its mean time between failures m, according to specification. Component failure rates for the estimated stress levels are obtained from component failure rate derating curves, or are supplied by the reliability organization in the company. Also, component on-times must be considered if they are different from the equipment's operating time. The techniques of reliability analysis have been explained in detail in the preceding chapters.

It is clear that in the preliminary reliability analysis stage only very rough estimates of component failure rates will be available or possible. As we have said before, this preliminary analysis serves comparative purposes to help decide which way to go in the final design.

The results of the preliminary analysis are used with advantage as recommendations for the final design stage. The numerical analysis of the chosen configuration shows where inherently weak areas may be expected, where component stresses should be reduced, and what quality of components should be used.

When the final design is completed, a phase 2 reliability analysis is made. This is no longer a purely theoretical affair, because the final configuration of circuits or layouts is now known and definite types of components have been chosen or designed. Much more exact circuit and stress analysis is now possible, and failure rate curves for definite component types can be requested from the component suppliers and used in the reliability analysis. Although this phase 2 reliability analysis is more important than phase 1 analysis, phase 1 cannot be omitted because it supplies the information which must be considered in the final design.

Phase 2 analysis often consists of two steps when circuit design and the design of packaging the components into a black box or assembly are made in two steps. The packaging usually determines the environmental stresses acting on the components, including temperature rise, and this substantially affects component failure rates. Also, maintainability must be considered in the design of the packaging arrangements and methods. Those components with the highest failure rates should be easiest to get at if the equipment is re-usable and subject to maintenance procedures. With all the design information available, including circuit and stress analysis, component types and their location, a good estimate of the equipment's reliability can be elaborated. If this estimate exceeds the specified reliability requirement with a reasonable margin, prototype production can begin. If the estimate does not exceed the required value, some changes in the design or packaging may be required to reduce the stresses acting on some components—better protection of such components, better heat sinks or cooling, and perhaps component rearrangement in the package, until the required reliability is analytically exceeded.

Taking the combined effects of phases 1 and 2 of the reliability control program, it is obvious that a number of important reliability achievements have been made during design, before any hardware has even had to be touched. The approach most promising from a reliability point of view was chosen, information was made available for the final design as to the areas where particular care had to be exercised not to overstress components or where redundancy must be applied, reliability was considered in packaging along with maintainability, if required, and properly rated components could be selected and specified. In this way an optimum design could emerge although the stresses expected to act on the individual components could be only very roughly estimated. Still, this is the best that can be done at the design stage, and if this reliability control program is intelligently followed through, very little trouble should occur in the later stages.

The actual physical stresses—electrical, mechanical, and thermal—will become known only when prototypes are built and tested. These tests are part of phase 3 of the reliability control program. They are in the nature of statistical reliability tests, and their purpose is to assure that the hardware reliability will be what the previous analyses have predicted. Factors overlooked during design and analysis may show up in these tests. The tests can be accelerated tests in which failure occurrence is forced; the failure occurrence is statistically evaluated and related to the previous reliability analyses. (Reliability testing methods will be discussed later in this book.)

As a result of phase 3 of the reliability control program, design changes in circuitry, layout, or packaging may be required, and sometimes components of higher ratings must be substituted. However, if phases 1 and 2 of the control program were properly carried out, only minor changes will be required, or none at all. Certainly complete redesign will be avoided, and no unreliable product will go into production to be scrapped when it proves to be a failure in operation.

Phases 1 and 2 assure that the design is proceeding on the right track, and phase 3 confirms this; if necessary, it irons out minor, overlooked shortcomings. Phase 3, without being preceded by phases 1 and 2, is, generally, of little or no effectiveness. The reliability of an equipment is decided by its design and cannot be worked into an equipment by testing.

The design determines the incidence and frequency of chance failures and the wearout life of an equipment, and therefore, also the incidence and frequency of wearout failures. Thus, the designer decides how often an equipment will fail in operation. This is especially true in the absence of quantitative reliability specifications spelled out by the customer. If reliability specifications are given, he is largely relieved of this responsibility. All he has to do, by contract, is to design the equipment so that its failure rate in operation will not exceed the specified requirement. Still, whenever possible, he should aim at designing the equipment to be more reliable than specified, especially when the equipment can cause loss of the system in which it is used as, for instance, in an airplane where loss of lives may also be involved, or when it is part of a weapon system or a weapon system itself, which may be a vital link in the security and preservation of the nation.

Thus the designer should be highly reliability-conscious and should never treat reliability lightly. It is not an additional nuisance—it is a most important part of his work. Naturally, he can follow this policy only if he gets full support from a reliability-conscious management which provides all the prerequisites for reliability work in the form of a supporting reliability organization capable of supplying the information on reliability data and methods needed and properly administering all stages of a reliability control program for each individual design effort.

Although the designer is the architect of reliability, he cannot do the job alone. He needs a supporting reliability organization which can provide the necessary data on components and, whenever needed, can perform for him probability calculations, especially when a project is under pressure to meet a deadline. Still, the designer must know enough about reliability because he synthetizes the equipment or the system from the components, which is decisive for reliability. He makes the final design decisions after weighing all the reliability aspects of a problem.

A crucial part of every reliability design is to obtain applicable failure rate data on components to be used. Such data is important for the selection of components and for the realism of reliability analyses. How to get reliable data is a frequent question, and sometimes a big problem.

In phase 1, during the preliminary reliability analysis when no definite types of components are yet decided upon, so-called *general failure rates* can be used, because the analysis is only comparative and serves to establish the relative reliability values of a number of possible design approaches and to find relatively weak reliability links in a certain design. Many companies have their own lists of average failure rates of the components used in their design work based on past experience which has been statistically evaluated. Such information is often quite helpful for comparative reliability analyses, especially where similar stress levels occur. It is good practice to compile such average failure rate data; they are derived by summing up the total accumulated component operating hours in service use and dividing this by the number of failures which occurred during that time.

For instance, if there are 10 components of a given type in an equipment and 20 such equipments were supplied to a customer a year ago and another 20 six months ago, and it is known that the equipment (installed in an aircraft, for instance) averages 300 operating hours monthly, one can evaluate the average failure rate of these components as follows: There are 200 components of the given type in the first 20 equipments. They have accumulated in 12 months $12 \times 300 \times 200 = 720,000$ unit hours. The components delivered later have accumulated $6 \times 300 \times 200 = 360,000$ unit hours. Together the total accumulated operating hours on the components are 1,080,000 unit hours. If in this entire period of one year the customer has reported 7 failures of the components involved, the average hourly failure rate of these components will be about $7:1,080,000 = 0.0000065$.

In this way a failure rate list can be compiled for various components of different types and manufacture. This applies mainly to components which are most frequently used in design work, and therefore, experience with these components is available either if it was noted how the equipment has behaved in service, or if information can be requested from its user who probably has logged failure occurrences (as do airlines). Such failure rate lists are very useful for reliability analysis work on similar equipment designed for similar use.

Also, there exists limited literature on component failure rates and derating curves, especially for electronic equipment.* From such curves the failure rates of the most frequently used electronic components can be directly obtained for various electrical and temperature stress levels.

Such data is useful for preliminary analysis work. It must be realized that the information is based on components of a definite type and manufacture. Those of another type or manufacture will display different failure rate characteristics.

Phase 2 analysis work requires a more exact approach. Of course, it is possible to obtain precise failure rate estimates only by testing very large

---

* See "Reliability Stress Analysis for Electronic Equipment," Technical Report No. 1100, Radio Corporation of America (October 1, 1956).

samples of a given component under exactly the same stress levels at which the component will operate in the equipment when put into service. Such large tests are not practical for the component user. However, the component manufacturer should test his components to know how good they are and what is he selling. Therefore, as soon as the types of components are chosen for the final design, failure rate data should be requested from the component manufacturers. In practice, this amounts to asking for the information defined by Equations (15.5) and (15.6) in Chapter 15, from which approximate failure rate curves can be drawn, or asking directly for valid failure rate derating curves. To be reliable, all such information, of course, must be supported by certified test reports. For the manufacturer of the components it is not an extraordinary effort to supply this information. Every component manufacturer is equipped for testing, and the problem is to set up regular reliability tests and test at three test points as suggested in Chapter 15; the information is then available to all customers. A good rule to remember is that it is safer to buy components from a manufacturer who makes reliability test information available than to buy components which have never been tested for reliability.

The minimum information which should be available on a component of a given type and make is its nominal failure rate when operated at its rated values of wattage, voltage, etc. If this is available, estimates for lower stresses can be made by applying the derating laws explained in Chapter 15.

In phase 2 reliability design review it also becomes necessary to estimate closely the stresses to which components will be exposed in the operating equipment. Electrical stresses can be estimated from electrical circuit analyses, thermal stresses from thermal analyses, etc. However, it is a difficult problem to estimate closely, for instance, the temperature at which a component will operate in a given location, including the temperature rise of the component if it is a heat dissipating device. Usually only rough estimates can be made, and these are used to derive component failure rates from the derating curves.

This is where phase 3 enters the reliability program, because during the simulated tests the components can be made to operate under stresses essentially equal to the operating stresses in service. The electrical parameters can be measured, temperature rise can be evaluated by the use of thermometers and thermocouples, and a review of component failure rates can then be made. Also, the incidence of failures during the tests, especially when accelerated testing is used, will be very revealing. Design changes, if necessary, can then be made before production is authorized. We then know that we are putting a mature, reliable design into production, and disappointments and financial loss can be avoided.

Another point which must be considered in electronic design is the parameter drift of components. This is actually a type of wearout phenomenon and means that some electronic components begin to wear out rather early, although the drift is comparatively slow. It usually takes a long time for a component to drift out of tolerance. The wider the acceptable tolerance limits, the longer a component will render useful service, or, in other words, the later an out-of-tolerance failure will occur. The design of tolerant electronic circuits takes advantage of this. Statistical design techniques, widely discussed in literature,* can be applied in the design of circuits tolerant to component value drifts.

The parameter drift phenomenon of certain components such as resistors, capacitors, etc. does not affect the chance failure rate of the components or of systems, and therefore does not directly affect the reliability of equipment—nor does this drift necessarily reduce the strength of the components involved. But it does affect the performance of electronic equipment, causing a gradual degradation of performance quality. This drift due to component operating age is usually permanent. Another phenomenon of component value drift occurs with temperature and is usually transient. When the temperature is restored to the previous level, the component also assumes its original parameters. Both these types of drift have basically the same effects on circuit performance. Therefore, when designing electronic circuits for performance, these drifts must be considered.

In designing for reliability, the designer has two powerful tools at his disposal—one is reliability analysis applied at all stages of the design, the other is the derating of components.

Reliability analysis is the calculation of the system's or equipment's reliability from the failure rates of the components used. The designer combines components to get the required performance, but he also must choose the most advantageous way of combining them to get the required reliability. By properly applying reliability analysis, the design can be broken up into modules, assemblies, units, or subsystems, and reliability apportionments can be made to these so that the over-all reliability target is met.

When some components are inherently not reliable enough, derating techniques must be used. The derating of components in reliability design is a standard technique—it means operating components at only one-half or even less of their rated values of voltage, wattage, temperature, etc. Spectacular increases in component and equipment reliability can be achieved in this way. Sometimes only one of these stress parameters needs to be reduced to bring the component failure rate down to the required level. But there are also instances where even extreme derating does not help, and the component failure rate is still too high. Redundancy techniques must then be used, in the form of parallel or stand-by components. However, redundancy increases size, weight, and possibly, maintenance as well. Therefore, every effort should be made to avoid redundancy if single components of high reliability can be obtained and when derating techniques can be used.

Because the reliability of electronic systems is mostly a series reliability with very little or no redundancy used, the derating of components becomes an absolute must for the achievement of high reliabilities. And because the reliability of a series system is determined by the sum of the component failure rates, design simplification is another important tool for reliable design. Simplification means a reduction of the number of components in an equipment. To achieve required performance with the minimum possible number of components is a sound approach to system reliability. Thus, the art of reliability design does not consist of designing highly sophisticated systems but rather of designing them to be simple as possible. It is definitely not a matter of showing how complicated and complex a system can be designed to give the required performance; rather, it is the art of designing the simplest system which will give satisfactory performance with a minimum chance of failure.

Summing up, the designer has certain tools for reliable design at his disposal. He can

1. Simplify the design to a minimum of parts without degrading performance.
2. Perform design reliability reviews by means of reliability analysis at the preliminary stage and at the final stage, and if necessary, in-between.
3. Apply component derating techniques to the best possible advantage to reduce failure rates and to increase component life.
4. Reduce the operating temperature of components in the equipment by providing heat sinks, appropriate packaging, and if necessary, good cooling.
5. Eliminate resonant vibrations by good isolation and protect equipment against shock, humidity, corrosion, etc. Use shock resistant packaging for missile application.
6. Specify component reliability and burn-in requirements.
7. Specify prototype tests and production equipment debugging procedures.

The specification of component reliabilities, component burn-in requirements, and production equipment debugging procedures is necessary to insure that the reliability achieved by the design will be maintained in production and will not be degraded by the infiltration of substandard specimens into the equipment or by poor assembly techniques which cause early failures. These requirements will alert incoming inspection, quality control, and production engineering to do their part in the over-all reliability control program.

Chapter 19

DESIGN ANALYSIS EXAMPLES

EXAMPLE 1. The radio communications system of an aircraft is required to have a reliability of 99.5 per cent for a 10-hour flight. The system is to consist of a receiver-transmitter, antenna control, antenna coupler, filter, and antenna, of specified performance.

A reliability block diagram of the basic system is shown in Figure 19.1. The failure rates in the first column are derived from past experience and from several sources. By comparing the environmental conditions in which the equipment is expected to operate in the proposed application with the conditions which yielded the historical data, we predict a new set of failure rates shown in column 2.

The expected total failure rate of the system adds up to 0.015 per hour, which amounts to a system reliability for 10 hours of

$$R_{10} = e^{-(0.015)(10)} = e^{-0.15} = 0.861$$

This is obviously short of the specified reliability value.

A look at the predicted failure rates shows that the predominant contribution to failure is made by the receiver-transmitter unit. Therefore, a re-evaluation of the reliability of this unit is made in the form of a preliminary general analysis, as shown in Figure 19.2. Because the analysis is of a preliminary character, the use of average failure rates of the components will yield sufficiently close approximations. Therefore, rather than performing a detailed analysis of each component, we apply the analysis to groups of components. This practice is often used for electronic equipment, especially in the tentative design stage, or when average data on similar equipment are available, as in this case. Information was available that the R-T unit under consideration was designed so that tubes operate on the average at 95 per cent of their full rating, whereas

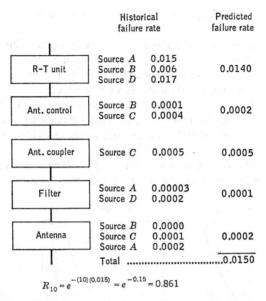

	Historical failure rate		Predicted failure rate
R-T unit	Source A Source B Source D	0.015 0.006 0.017	0.0140
Ant. control	Source B Source C	0.0001 0.0004	0.0002
Ant. coupler	Source C	0.0005	0.0005
Filter	Source A Source D	0.00003 0.0002	0.0001
Antenna	Source B Source C Source A	0.0000 0.0001 0.0002	0.0002
	Total		0.0150

$$R_{10} = e^{-(10)(0.015)} = e^{-0.15} = 0.861$$

Fig. 19.1. Block diagram of a basic radio communications system.

	Failure rate* per component	Failure rate* per group	Derated failure rate per component	Derated failure rate per group
73 tubes	0.120	8.760	0.090	6.5700
617 resistors	0.001	0.617	0.0008	0.4936
584 capacitors	0.002	1.168	0.0003	0.1752
24 relays	0.035	0.840	0.0350	0.8400
51 transformers and coils	0.010	0.510	0.0100	0.5100
11 switches	0.008	0.088	0.0080	0.0880
Total failure rate		11.983		8.6768

Original ⟋ Derated ⟋

*Failure rate per 1000 hours

Fig. 19.2. Simplified reliability analysis of a radio transmitter.

all other components operate on the average at 90 per cent of their full rating.

From a specific set of component derating curves shown in Figure 19.3 which apply to the type and make of components used and to the specified ambient operating temperature of 60°C, the average failure rates of components were derived for the given stress levels. These numbers are given in the first column of Figure 19.2 and are summed up for the entire R-T unit in the second column. The total amounts to a failure rate of approximately 12 per 1000 hours, or 0.012 per hour. This agrees well with the prediction of 0.014 made in Figure 19.1.

To make the R-T unit more reliable, its redesign by means of a conservative component re-rating is considered, without sacrificing perform-

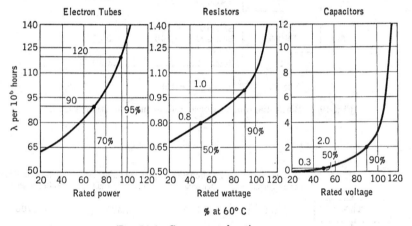

Fig. 19.3. Component derating curves.

ance. Some components of higher ratings may have to be selected and operated at relatively lower stress levels.

The third column in Figure 19.2 shows the new expected average failure rates with tubes operated at 70 per cent of maximum rated power, resistors at 50 per cent rated wattage, and capacitors at 50 per cent rated voltage. These failure rates are also derived from Figure 19.3, as shown.

Summing up the new failure rates in column 4 of Figure 19.2, we obtain a total failure rate of approximately 0.009 per hour for the R-T unit. This represents a 25 per cent failure-rate reduction. Recalculating the predicted reliability with the new failure rate, we obtain a reliability for the whole system of 0.905, or 90.5 per cent. This is a definite improvement compared with the previous value of 0.861.

But this reliability is still far from the required goal of 99.5 per cent. Therefore, we must consider redundant operation of two or more units in parallel in various configurations, shown in block diagram form in Figure 19.4. The upper portions of the reliability curves of the five configurations are shown on the left in the same figure. The equations from which these curves are derived are as follows:

(A) $R = R_{RT}R_A$

(B) $R = (1 - Q_{RT}^2)R_A$

(C) $R = 1 - (1 - R_{RT}R_A)^2$

(D) $R = (1 - Q_{RT}^2)R_{SW}(1 - Q_A^2)$

(E) $R = (1 - Q_{RT}^3)R_{SW}(1 - Q_A^2)$

The subscript RT stands for the receiver-transmitter, A for the antenna

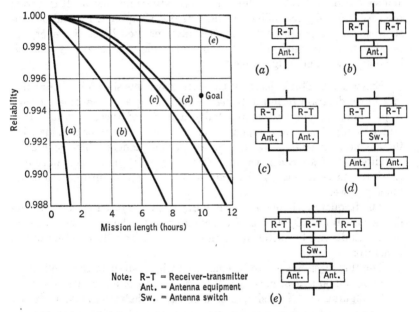

Note: R-T = Receiver-transmitter
Ant. = Antenna equipment
Sw. = Antenna switch

Fig. 19.4. Reliability comparison of radio communication system configurations.

equipment including control coupler and filter. R and Q denote reliability and unreliability, respectively.

Configurations C, D, and E have two antennas, which in D and E can be switched to either receiver-transmitter.

A study of the reliability curves shows that of the discussed configurations, only E meets the reliability specification of a minimum of 0.995 for 10 hours.

In practice, however, configuration E would not be applied because of weight and size considerations with the three-unit system. Therefore, efforts would be made to increase the reliability of configurations C or D until the system reliability goal is met. A reliability review of the antenna equipment similar to the R-T unit review, additional derating, cooling the equipment by blowers, or the provision of heat sinks would achieve the purpose. The reliability curves of these two configurations indicate that only moderate reliabilization efforts will be needed to upgrade reliability to the required level.

The calculations in this example were made on the assumption that all parts of the system are kept in an "on" condition for the full 10 hours of mission duration, but this is not necessarily the case. Some units may be used in stand-by redundancy, being kept inactive in an "off" condition and switched on only if a failure in the operating units occur. If this were the case, stand-by reliability calculations (see Chapter 12) would be substituted for the parallel reliability calculations used here. However, the parallel case discussed above is a good example of a reliability analysis in an early system design stage.*

EXAMPLE 2. Three electrical generating systems shown in Figure 19.5 in a block-diagram form are being considered for a twin-engine aircraft. A preliminary comparative reliability analysis of these systems is required. Only the major components of the systems are to be considered— the engines, the generators, and the frequency changers. The generators are direct, engine-driven, variable frequency machines. Their output is fed into static frequency changers which convert it into constant-frequency power.

The frequency changers in configuration A are rated 60 kva, in configurations B and C, 30 kva. In configuration C, the frequency changer in the middle can be switched automatically or manually to either of two generators.

The three systems are to be compared with respect to their reliability of supplying 60 kva of normal power and 30 kva of emergency power.

Configuration A is a simple case of two paths in parallel. Only one path can be allowed to fail if normal operation at 60 kva or emergency operation at 30 kva is to be maintained. Therefore, the system reliability of supplying 60 kva is identical with its reliability for emergency operation:

$$R_{60} = R_{30} = 1 - (1 - R_E R_G R_F)^2$$

where R_E, R_G, and R_F are the reliabilities of the engine, generator, and frequency changer respectively.

Configuration B has two generators on each engine and is capable of supplying 30 kva emergency power if at least one of the two engines and

---

* This example of reliability analysis was supplied by Mr. Roger Florence, reliability engineer with the Boeing Airplane Company at Renton, Wash.

at least one of the two generator-frequency changer sets driven by the surviving engine operates. Therefore,

$$R_{30} = 1 - \{1 - R_E[1 - (1 - R_G R_F)^2]\}^2$$

The system's reliability of supplying 60 kva can be calculated by a step-by-step application of Bayes' theorem for the system's probability of

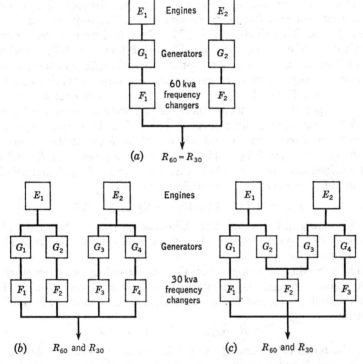

Fig. 19.5. Twin-engine aircraft generating system configurations.

failure Q_{60}. For instance, starting with the generator-frequency changer set on the left, which has a reliability $R_{S_1} = R_{G_1} R_{F_1}$ and an unreliability $Q_{S_1} = 1 - R_{G_1} R_{F_1}$ we write the system's unreliability as follows:

$$Q_{60} = R_{S_1} Q' + Q_{S_1} Q''$$

where the probability of system failure if set S_1 is good equals

$$Q' = R_{E_1}(1 - R_{S_2})\{1 - R_{E_2}[1 - (1 - R_{S_3})(1 - R_{S_4})]\} \\ + (1 - R_{E_1})(1 - R_{E_2} R_{S_3} R_{S_4})$$

and probability of system failure if set S_1 is bad equals

$$Q'' = R_{E_1}R_{S_2}\{1 - R_{E_2}[1 - (1 - R_{S_3})(1 - R_{S_4})]\}$$
$$+ (1 - R_{E_1}R_{S_2})(1 - R_{E_2}R_{S_3}R_{S_4})$$

Equating the terms $R_{E_1} = R_{E_2}$ and $R_{S_1} = R_{S_2} = R_{S_3} = R_{S_4}$, we obtain system reliability from $R_{60} = 1 - Q_{60}$ as

$$R_{60} = (2R_E + 4R_E^2)R_G^2R_F^2 - 8R_E^2R_G^3R_F^3 + 3R_E^2R_G^4R_F^4$$

The same result can also be obtained by expanding the reliability binomial of the four generator-frequency changer sets so that any two out of four must operate. This yields $6R_S^2 - 8R_S^3 + 3R_S^4$, and we consider for each term the ways in which it occurs. Thus, the first term $6R_S^2$, which requires that at least two generator-frequency changer sets survive, can occur in six ways. Two ways occur when sets 1 and 2 survive or sets 3 and 4 survive. For each of these the respective engine must also survive so that we get $2R_ER_S^2$. Further, there are four ways in which sets 1 and 3, 2 and 3, 1 and 4, or 2 and 4 survive. For each of these four ways both engines must survive, and we get $4R_E^2R_S^2$. Thus, together the six ways in which R_S^2 occurs are $(2R_E + 4R_E^2)R_S^2$. For the occurrence of R_S^3 and R_S^4 both engines must survive in both cases, and we get $8R_E^2R_S^3$ and $3R_E^2R_S^4$. The final formula for system reliability then reads

$$R_{60} = (2R_E + 4R_E^2)R_S^2 - 8R_E^2R_S^3 + 3R_E^2R_S^4$$

This approach by means of the binomial expansion is faster than the approach by Bayes' theorem when equal components are involved in parallel operation.

The reliability of configuration C, which also has two generators on each engine, can again be calculated by means of Bayes' theorem. For both reliabilities R_{30} and R_{60} we use the same basic system unreliability equation:

$$Q = R_{G_1}R_{F_1}Q' + (1 - R_{G_1}R_{F_1})Q''$$

Only the terms for Q' and Q'' will differ for the 60-kva and 30-kva requirements. For 60 kva, Q' is derived as follows:

$$Q'_{60} = R_{E_1}\{R_{G_2}(1 - R_{F_2})(1 - R_{E_2}R_{G_4}R_{F_3})$$
$$+ (1 - R_{G_2})[1 - R_{E_2}(R_{G_4}R_{F_3} + R_{G_3}R_{F_2} - R_{G_3}R_{G_4}R_{F_2}R_{F_3})]\}$$
$$+ (1 - R_{E_1})(1 - R_{E_2}R_{G_3}R_{G_4}R_{F_2}R_{F_3})$$

and

$$Q''_{60} = R_{E_1}R_{G_2}(1 - R_{E_2}R_{G_4}R_{F_2}R_{F_3})$$
$$+ (1 - R_{E_1}R_{G_2})(1 - R_{E_2}R_{G_3}R_{G_4}R_{F_2}R_{F_3})$$

The terms for Q' and Q'' are based on the requirement that at least two out of the three frequency changers must operate, with their respective generators and engines, to obtain 60 kva. The final formula for R_{60}, after equating the terms $R_{E_1} = R_{E_2}$, etc., reads:

$$R_{60} = R_F^2(3R_E^2R_G^2 - 2R_E^2R_G^3 + 2R_ER_G^2) - R_F^3(4R_E^2R_G^3 - 2R_E^2R_G^4)$$

For 30 kva the terms Q' and Q'' must be based on the requirement that at least one out of three frequency changers must put out 30 kva:

$$Q'_{30} = (1 - R_E)[1 - R_E(2R_GR_F - R_G^2R_F^2)]$$

$$Q''_{30} = R_ER_G(1 - R_F)(1 - R_ER_GR_F)$$
$$+ (1 - R_ER_G)[1 - R_E(2R_GR_F - R_G^2R_F^2)]$$

The resulting reliability of the system to supply 30 kva is then

$$R_{30} = 1 - R_GR_FQ'_{30} + (1 - R_GR_F)Q''_{30}$$

Assuming constant failure rates for the jet engines (0.00005), the inductor generators (0.0001), and the frequency changers (0.0005), R_E, R_G, and R_F can be calculated for various operating periods and the reliabilities R_{60} and R_{30} of the three systems can be evaluated and compared with each other.

Configuration B is the most reliable, but configuration C may be acceptable according to what reliability and safety requirements have been specified. The reliability of configuration C can be further increased if automatic or manual switchover is provided so that any frequency changer can be operated from any of the four induction generators. When the frequency changers are located in the fuselage, such switching arrangements become practical. Automatic switchover can be backed up by manual switching, in which case the reliability of switching becomes unity for all practical purposes, because the switching is a one-cycle operation only. In three-engine airplanes only three generators would be used, but a switchover of the frequency changers would still provide sufficient safety for an emergency electric power supply.

EXAMPLE 3. The reliability of supplying hydraulic pressure to the junction point of the brake system of a twin-engine aircraft is to be investigated. Propeller-type engines are used, and each powers a hydraulic pump which together supply pressure to the aircraft's hydraulic utility system. Each pump is capable of supplying the fluid volume and pressure required, and therefore the pumps operate in parallel. All connections and fittings in the utility hydraulic pressure system constitute potential leakage points and therefore appear as series elements in the block diagram shown in Figure 19.6.

The reliability R_A of volume and pressure being available at the junction point from which the left and right brake systems are powered amounts to

$$R_A = R_2[1 - (1 - R_1)^2] = e^{-0.00423}[1 - (1 - e^{-0.006})^2] = 0.9957$$

for a flight of 10 hours. This would mean that in an average of about four or five landings out of 1000 such flights, no normal pressure for breaking would be available.

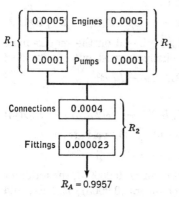

If the aircraft also has an auxiliary hydraulic system powered by an electric motor-driven pump, a tie-in of this auxiliary system to the brake junction point may increase the reliability of normal braking by two orders of magnitude. Check valves can be used to prevent the utility system and the auxiliary system from interfering with each other.

Fig. 19.6. Hydraulic pressure supply.

Such tie-in of a hypothetical auxiliary system is shown in the block diagram of Figure 19.7. It results in an improvement of the probability that normal pressure will be available at the junction point from $R_A = 0.9957$ to

$$R = 1 - (1 - R_A)(1 - R_B) = 1 - 0.0043 \times 0.0052 = 0.99997$$

and normal pressure, after the tie-in, would therefore be lacking in an

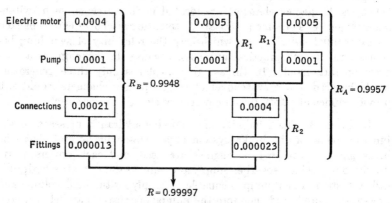

$R = 0.99997$

R_A = Reliability of the utility hydraulic system
R_B = Reliability of the auxiliary hydraulic system
R = Probability of pressure at junction point

Fig. 19.7. Hydraulic pressure supply with auxiliary system tie-in.

average of only about three landings out of 100,000 flights. Thus, parallel redundancy would increase this reliability by about two orders of magnitude.

Chapter 20

OVERHAULS AND
PART REPLACEMENTS

I N CHAPTER SEVENTEEN we discussed maintenance policies and derived the system utilization factor, availability, and dependability considering the manhours and the clock time required to restore system operation.

The repair time required for each component varies according to its physical characteristics and location in the system, but variations of the repair time also exist for identical components in identical locations. There is a certain minimum time required to get to the faulty component and to remove it as well as to put it back again or to replace it by another. This time varies with skill, as does the actual repair time, but it is safe to assume that for a given component in a given location, the over-all repair time is approximately normally distributed, with some mean and standard deviation. When the mean and the standard deviation are experimentally estimated for a given crew over a sufficiently large number of repairs performed on a particular component in a particular system, the normal distribution curve of the repair times can be derived, and from this can be derived the probability that the repair will be performed in a given time.

Obviously, from the normal curve the probability that the repair will be performed in a time equal to the mean time M is 0.5 or 50 per cent, and the probability that the repair will be completed in a time of, say, $M - 3\sigma$ is only 0.0014 or 0.14 per cent, etc. This probability of repairing a particular component in a given system in a given time is sometimes referred to as the *maintainability* of that component. However, as we have shown in Chapter 17, it is possible and more meaningful to use the mean or average repair time for components in manhours, convert this to clock time, and then derive the average system down time and system availability as was shown.

In system availability calculations, the down time required by both scheduled overhauls and off-schedule repairs must be used. In system dependability calculations, only the down time required by off-schedule repairs is considered because we have defined dependability as the probability or the per cent of time that the system can be expected to perform between two regular overhauls.

The functional down time between two regular or scheduled overhauls depends on system reliability which determines the average number of system failures and on the time it takes to repair these failures.

When the mean time between failures of a system is m, conclusions can be drawn as to how often the system will fail in a time T_o between two regular overhauls, where T_o is the system's operating time and amounts to the calendar time T less the total down time T_D.

If between two regular overhauls the system is not affected by wearout failures so that it behaves exponentially, its reliability of operating for T_o hours between the two overhauls is

$$R(T_o) = e^{-T_o/m} \tag{20.1}$$

where T_o is the required operating time.

For instance, if the overhaul time of an equipment or system is fixed for every $T_o = 800$ hours of operation and the mean time between failures of this equipment is $m = 4000$ hours, the probability that the equipment will operate for 800 hours without failing is

$$R = e^{-T_o/m} = e^{-800/4000} = e^{-0.2} = 0.81873$$

This means that we would expect about 82 per cent of these equipments to reach the regular overhaul time without failing, and about 18 per cent would fail before the time T_o. Therefore, about 18 per cent of the equipments would have to be overhauled before the regular 800 hours. Thus the expected average overhaul time for all these equipments would be less than 800 hours.

The average time between overhauls T_{avg} is computed as a mean time for the operating period T_o, similar to the way the mean time between failures is computed from R as the integral of R from zero to infinity. The time T_{avg} is actually a mean time between both scheduled and unscheduled overhauls:

$$T_{avg} = \int_0^{T_o} R \, dt \tag{20.2}$$

In the exponential case this equals

$$T_{avg} = \int_0^{T_o} e^{-t/m} \, dt = -m \left[e^{-t/m} \right]_0^{T_o}$$
$$= -m(e^{-T_o/m} - 1) = m(1 - e^{-T_o/m}) = mQ(T_o) \tag{20.3}$$

thus, in general, T_{avg} equals the mean time between failures multiplied with the unreliability for the overhaul period.*

In our case, with $Q(T_o) = 0.18$ and $m = 4000$,

$$T_{avg} = 4000 \times 0.18 = 720 \text{ hours}$$

Thus, if we had 100 of these equipments with a scheduled time between regular overhauls of $T_o = 800$ hours, about 82 would operate without failure for the full 800 hours, about 18 would fail and would have to be repaired before 800 hours, and averaging over the 100 equipments, the average time between overhauls would be 720 hours. The knowledge of this average time allows us to plan for a total number of overhauls over a long period of time.

If the overhaul time were extended from 800 hours to 1000 hours, and if we could assume that wearout will not affect such extended operation, the effect of the extension is calculated as follows:

$$R(T_o) = e^{-1000/4000} = 0.7788$$

which means that now only about 78 per cent of the equipment will reach the scheduled overhaul without failing, and about 22 per cent will fail earlier. If the policy is to overhaul an equipment when it fails, and not merely to repair the specific failure (as is done with major equipment such as aircraft engines, etc.), then the average time between overhauls would be about

$$T_{avg} = m(1 - 0.78) = 4000 \times 0.22 = 880 \text{ hours}$$

as against the scheduled 1000 hours. Thus, the percentage of equipment failing between regular overhauls has increased from 18 to 22, but the total number of overhauls over a long period of time has been reduced. We should mention that this extension of the overhaul period does not affect the reliability of the equipment as long as we are sure that wearout failures will not occur.

On the other hand, the main function of overhaul is to prevent wearout. In Chapters 6 and 7 we have shown how long components and equipments can be allowed to operate so as to minimize the effect of wearout. For single components the replacement or overhaul time must be kept at $M - 3\sigma$ or $M - 4\sigma$ or in between to prevent wearout from appreciably increasing the failure rate. If large numbers of components are in a system, this replacement or overhaul time must be further reduced—to $M - 5\sigma$ or even $M - 6\sigma$, according to the reliability requirements—which means according to what increase in system failure rate over the chance failure rate can be tolerated without reducing system reliability by wearout failures below the specified level.

* Igor Bazovsky, "Boeing Airplane Company's Reliability Handbook," Document D6-2770 (1957), Part I, pp. 12 and 13.

Assume that the equipment in our example, which has a mean time between failures of $m = 4000$ hours, has a mean wearout life of 1200 hours with a standard deviation of $\sigma = 100$ hours. With an overhaul period fixed to $T_o = 800$ hours, the probability of wearout failure occurring at or prior to the 800 hours is that corresponding to $M - 4\sigma$ and amounts to 0.0000317, which is negligible compared to the probability of chance failure of 0.18. When the overhaul period is extended to 1000 hours, the probability of wearout failure increases to 0.0228 corresponding to $M - 2\sigma$. Therefore, the total probability of failure in 1000 hours increases from 0.22 to about 0.2428, with a corresponding decrease of system reliability for 1000 hours from 0.78 to 0.757, approximately—a noticeable reliability reduction. If the overhaul period were extended further, to, say 1100 hours, the percentage of equipments failing at or before 1100 hours would become about 38 per cent.

An extension of the time between overhauls is thus a question of the equipment's mean wearout life and of its standard deviation. Such extension, in order not to affect equipment reliability, requires that the mean wearout life of the equipment also be increased; otherwise, the reliability of the equipment may be dangerously reduced, especially during the operating hours just prior to overhaul.

Sometimes it is required that not more than x per cent of equipment fail prior to the scheduled overhaul period of T_o hours. In designing such equipment both its chance failure and wearout failure rates must be considered. The required equipment reliability for T_o hours is then $R(T_o) = 100 - x$ per cent and is the product of $R_C(T_o)R_w(T_o)$, i.e., of the probabilities that no chance failure and no wearout failure will occur. These probabilities can be varied in several ways to give the same product; therefore, various values can be chosen for m, M, and σ as design goals, as long as the proportion of chance failures to wearout failures is not specified. This proportion, however, decides the reliability of the system for short operating stretches immediately prior to and after an overhaul. The rapid increase of the wearout failure rate in the vicinity of the mean wearout life was shown in Figure 6.5, Chapter 6.

If we assume that an equipment is subject to wearout failures only and not chance failures, which is often the case, then the requirement for not more than x per cent failing between two regular overhauls or between two scheduled replacements determines the $T_o = M - n\sigma$ value of the normal density function. From normal tables n can be evaluated, and if previous experience gives a clue to the magnitude of σ for similar equipment (for instance, $\sigma = M/10$), the design goal for the equipment's life is calculated as

$$M = T_o + n\sigma \qquad (20.4)$$

where n is read directly from the normal tables, σ is an experience value, and T_o is given together with the requirement of not more than x per cent failing prior to T_o. If x, for instance, is specified not to exceed 1 per cent, then $n = 2.34$, from normal tables, and for $T_o = 1000$ hours and an experience value of $\sigma = M/10$, the design life will have to be at least or more than

$$M = 1000 + 234 = 1234 \text{ hours}$$

which is then the equipment's mean wearout life.

Similar considerations also apply to parts replacements. If parts are regularly replaced to minimize the occurrence of wearout failures, the case becomes identical with that of regular overhauls. Again chance and wearout combinations must be considered if both types of failures are known to occur. If the chance failure rate is assumed to be zero, then we only need consider wearout. If the regular replacement time T_o is such that wearout is practically eliminated, then only chance failures are considered.

In practice, however, especially with electronic equipment, components are often allowed to operate until they fail and cause the system to malfunction. As we have seen in Chapter 7, such a system assumes after a stabilization time a comparatively very high failure rate which is derived from the replacement rate of the components, and wearout plays a predominant role when this policy of replacing components only as they fail is adopted instead of a policy of regular overhaul and part replacement schedules. This is obvious when we consider a component which has a mean time between (chance) failures of, say, $m = 1{,}000{,}000$ hours but its mean wearout life is only, say, 10,000 hours. After one year of continuous operation the component will have a high probability of failing because of wearout. The probability of its having failed because of chance in that year, however, is very small—approximately

$$\frac{t}{m} = \frac{8760}{1{,}000{,}000} = 0.00876$$

and its probability of surviving chance in 8760 hours is thus 0.99124; for 10,000 hours it is about 0.99. But its chance of surviving wearout for 10,000 hours is only 0.5. Thus it is clear that in the long run there would be only about one chance failure for every 50 wearout failures, or a ratio of $2:100$.

On the other hand, if regular part replacement schedules are adopted, wearout failures can be substantially reduced or, for all practical purposes, even eliminated by a proper choice of the replacement time $T_o = M - n\sigma$. Then only chance failures would occur and the probability that the equipment would fail in operation is drastically reduced. When only chance failures can occur between regular replacements, the number of parts of the same kind which will have to be replaced because of failing prior to the regular replacement time T_o amounts, on the average, to

$$K = NQ(T_o) \qquad (20.5)$$

where N is the number of like parts in the equipment and $Q(T_o)$ is the unreliability of one part for an operating time T_o between two regular replacements. When $Q(T_o)$ is small and only chance failures occur, the approximation $Q(T_o) = T_o/m$ can be used and we obtain

$$K = \frac{NT_o}{m} \qquad (20.6)$$

This formula can be readily evaluated for each category of parts m in an equipment or system.*

Now let us consider the effects of a given maintenance policy on the reliability of redundant systems. In Chapters 11, 12, and 13 we learned how to derive reliability equations for parallel, stand-by, and more complex systems. We have also calculated the mean time between failures, m, of these systems by integrating their reliability function $R(t)$ from $t = 0$ to infinity. Thus, on the average such redundant systems would fail once every m hours if failed redundant components were not replaced until system failure. However, if a maintenance policy is adopted which allows for the replacement or repair of failed redundant components before the system fails, system failure can be postponed depending on how often the system is inspected, and maintained if inspection reveals the presence of failed components. With this maintenance policy the system will fail less frequently than it would without inspections because it is assured that every new operating period after an inspection begins with full redundancy restored. The mean time between system failures thus becomes longer than m, and theoretically it would become infinitely long if failed redundant components were immediately replaced.

To find the correlation between T, by which we denote the operating time between two inspections, and m_T, which is the mean time between failures adopted by the system if it is inspected every T hours, we go back to Equations (20.2) and (20.3). The time T_{avg} in those equations represents the average time between scheduled and unscheduled replacements. It is thus a mean time at which the system is restored to its original condition. However, of these system renewal actions, only $Q(T) \times 100$ per cent are caused by system failure, whereas $(1 - Q(T)) \times 100$ per

---

* The treatment of parts replacements when wearout is included can be found in the article, "Two-Parameter Lifetime Distributions for Reliability Studies of Renewal Processes," by B. J. Flehinger and P. A. Lewis, *IBM Journal* (January 1959), pp. 58–73. See also the third example at the end of Chapter 22, where the number of replacement parts is estimated for a predetermined assurance level.

cent or $R(T) \times 100$ per cent are carried out for the purpose of preventing wearout without system failures having occurred. Making use of these equations, we can mathematically redefine the mean time between failures of a renewable device or system in terms of the time T between renewals:

$$m_T = \frac{T_{\text{avg}}}{Q(T)} = \frac{\int_0^T R(t)\, dt}{Q(T)} \tag{20.7}$$

which is the ratio of the expected or average time T_{avg} between scheduled and unscheduled renewals to the fraction of renewals caused by actual failure of the system. This is also consistent with the definition of the probability of failure $Q(T)$, which is the ratio of the mean time between all renewals (scheduled and unscheduled) to the mean time between failures (i.e. between the renewals caused by actual failure):

$$Q(T) = \frac{\int_0^T R(t)\, dt}{m_T} \tag{20.8}$$

In the case of an exponential series system the ratio in Equation (20.7) is always constant, regardless of T, and amounts to

$$m_T = m = \int_0^\infty R(t)\, dt$$

as long as the system components remain exponential—that is, as long as no wearout effects appear or as long as the components are preventively replaced before wearout can affect them. But in a nonexponential system, for instance, in a redundant system or if components suffer wearout, m_T the mean time between failures becomes a function of the scheduled renewal time T.

However, if the renewal time T is regular, even a nonexponential system will have the same probability of failure for each interval T between two scheduled renewals, overhauls, or inspections which ensure that the system begins each period T in its original good condition, and therefore over long periods of time will fail exponentially with a mean time between failures m_T given by Equation (20.7), and with an average constant failure rate given by

$$\lambda_{\text{avg}} = \frac{1}{m_T} = \frac{Q(T)}{\int_0^T R(t)\, dt} \tag{20.9}$$

For a regularly inspected redundant system, the scheduled renewal time T is the time between two regular inspections. If an inspection reveals failed redundant components, they are replaced or overhauled so that the system redundancy is fully restored. If the inspection does not reveal any failures, the system is considered *eo ipso* renewed as long as its components behave exponentially and because full redundancy is present in the system. The mean time between failures m_T and the average failure rate λ_{avg} are then computed from Equations (20.7) and (20.9) by substituting for T the time between scheduled regular inspections.

Figure 20.1 shows the mean time between failures m_T of a regularly

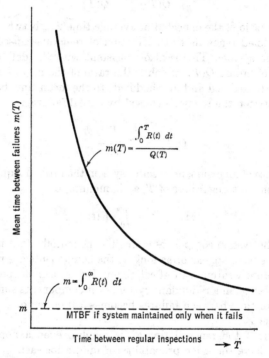

Fig. 20.1. Mean time between failures of a regularly inspected and maintained redundant system.

inspected redundant system as a function of the inspection time T. The more frequently the system is inspected, that is, the shorter is T, the longer will be the system's mean time between failures m_T. Conversely, the longer T is made, the shorter becomes its mean time between failures, and in the limit, when the system is not inspected regularly at all, i.e., when $T = $ infinity, the system's mean time between failures becomes

$$ m = \int_0^\infty R(t)\ dt $$

This follows from Equation (20.7) because $Q(T) = 1$ for $T = $ infinity.

For a better understanding of the concept of a redundant system's average failure rate λ_{avg}, Figure 20.2 shows the curves of the instantaneous failure rate

$$\lambda(t) = -\frac{1}{R}\frac{dR}{dt} = -\frac{d(\ln R)}{dt}$$

as a function of the operating time t, and of $\lambda_{avg}(T)$ as a function of the inspection time T.

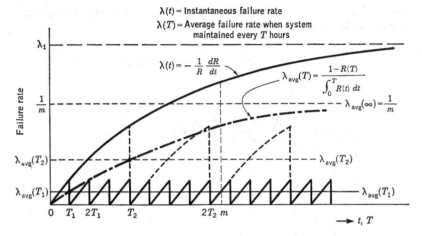

Fig. 20.2. Failure rate of redundant system without and with preventive maintenance.

If the system is not inspected, $\lambda(t)$ approaches in the limit the failure rate λ_1 of the system's most reliable redundant element which is capable of performing the system function alone. But if the system is inspected regularly every T_1 hours, its instantaneous failure rate $\lambda(t)$ drops back to zero every T_1 hours as the system is renewed, which is represented by the zigzag line at the bottom of the graph, and the system adopts a constant average failure rate of $\lambda_{avg}(T_1)$ which can be calculated from Equation (20.9) by substituting T_1 for T. And if the system were inspected only every T_2 hours, it would adopt the constant failure rate $\lambda_{avg}(T_2)$, as shown in the graph. The dash-point curve shows the average failure rate of the system as a function of the inspection time T.

It is important to realize that the directly observable quantity in the behavior of a regularly maintained redundant system is its average failure rate. Therefore, λ_{avg} can be added to the failure rates of any other components which operate in series with the redundant group. This addition process is shown in Figure 20.3, where a redundant group, inspected every

T hours, operates in series with a number of components the total of whose failure rates is λ_0. The system failure rate is then

$$\lambda_s = \lambda_0 + \lambda_{\mathrm{avg}}(T)$$

and system reliability becomes an exponential function:

$$R_s = \exp\,(-\lambda_s t)$$

From Figures 20.2 and 20.3 we can clearly see that the failure rate λ_{avg} which a redundant system actually assumes in service is determined by our choice of the inspection time T, and this also determines the system's reliability in actual service.

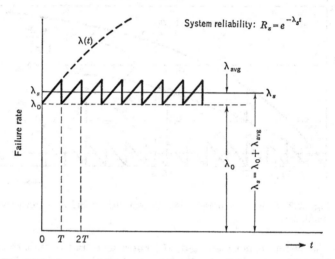

Fig. 20.3. Average failure rate of a series-parallel system. (Regularly inspected and maintained at T.)

As to the instantaneous failure rate $\lambda(t)$ of the system, this is a probability concept. We can ask at the time $t = 0$ when the system goes into operation with full redundancy built into it what its instantaneous failure rate will be at any arbitrary time t if we do not maintain the system in between. The curve $\lambda(t)$ shows this function. But if out of curiosity we were to inspect the system at a time t, we would find that the redundant system has either a zero failure rate if at least two redundant elements in the system are still operative, or it has a constant failure rate λ_1 if only one of the redundant elements is still alive. The system simply cannot have any failure rate other than zero or λ_1, and therefore the advance evaluation of $\lambda(t)$ is a theoretical matter which never complies with the true condition of the system. Thus, at the time t, a redundant system will be in a state with either a zero failure rate or a λ_1 failure rate, or it will be failed. Now the meaning of $\lambda(t)$ is that over a large number of observations the three possible system conditions would average out as if at time t the system had the theoretical instantaneous failure rate $\lambda(t)$. Therefore, $\lambda(t)$ allows us to calculate the system's cumulative probability of surviving from $t = 0$ to t, i.e., the system's reliability if the system is not maintained in between:

$$R(t) = \exp\left(-\int_0^t \lambda(t)\, dt\right) \qquad (20.10)$$

If the system survives up to the time t_1, and at t_1 we do not inspect the system so that we do not know whether or not redundant elements have failed, and therefore we do not know whether the system is in a zero failure rate or a λ_1 failure rate condition, its reliability of operating from t_1 to a subsequent time t_2 is given by

$$R = R(t_2 - t_1) = \exp\left(-\int_{t_1}^{t_2} \lambda(t)\, dt\right) \qquad (20.11)$$

But this will not be the system's true reliability of operating from t_1 to t_2. Its true reliability can be determined only by inspecting the system at t_1 to determine how much redundancy, if any, is still contained in the system and by calculating system reliability accordingly. However, over long periods of observation we would find that the system's reliability averages out to the value given by Equation (20.11). This equation can also be written with logarithmic exponents when making use of the equality of

$$\int_0^t \lambda(t)\, dt = -\ln R(t)$$

which in turn can be approximated by $Q(t)$ when $R(t)$ is larger than 0.9:

$$R(t_2 - t_1) = \exp\left[\ln R(t_2) - \ln R(t_1)\right] \cong \exp\left[Q(t_1) - Q(t_2)\right] \qquad (20.12)$$

which facilitates numerical evaluation once the survival characteristic $R(t)$ of a redundant system has been derived.

An approximation is sometimes also used for the evaluation of the average failure rate of redundant systems regularly inspected every T hours, the exact value of which is given by (20.9). This approximation reads

$$\lambda_{\text{avg}}(T) = \frac{-\ln R(T)}{T} \qquad (20.13)$$

and is based on averaging the area under the $\lambda(t)$ curve between zero and T. This procedure is not correct but it can be used as long as T is much smaller than the system's mean time between failures without preventive maintenance m.

In the following we give the equations for some frequently occurring redundant arrangements:

1. Two equal components in standby, with λ failure rate each:

$$R(t) = e^{-\lambda t}(1 + \lambda t)$$

$$\lambda(t) = \frac{\lambda^2 t}{1 + \lambda t} = -\frac{1}{R(t)} \cdot \frac{dR(t)}{dt}$$

$$m = \frac{2}{\lambda} = \int_0^\infty R(t)\, dt$$

$$\int_0^T R(t)\, dt = \frac{2}{\lambda} - e^{-\lambda T}\left(\frac{2}{\lambda} + T\right) = m - e^{-\lambda T}(m + T)$$

$$m_T = \frac{\int_0^T R(t)\, dt}{1 - R(T)} = \frac{m - e^{-\lambda T}(m + T)}{1 - e^{-\lambda T}(1 + \lambda T)}$$

$$\lambda_{\text{avg}}(T) = \frac{1}{m_T}$$

2. Two equal components in parallel, with λ failure rate each:

$$R(t) = e^{-\lambda t}(2 - e^{-\lambda t})$$

$$\lambda(t) = 2\lambda \frac{1 - e^{-\lambda t}}{2 - e^{-\lambda t}}$$

$$m = \frac{3}{2\lambda}$$

$$m_T = \frac{m - e^{-\lambda T}(2/\lambda - e^{-\lambda T}/2\lambda)}{1 - e^{-\lambda T}(2 - e^{-\lambda T})}$$

$$\lambda_{\text{avg}}(T) = \frac{1}{m_T}$$

3. Three equal components in parallel, with λ failure rate each, and at least two must operate to complete mission:

$$R(t) = e^{-2\lambda t}(3 - 2e^{-\lambda t})$$

$$\lambda(t) = 6\lambda \frac{1 - e^{-\lambda t}}{3 - 2e^{-\lambda t}}$$

$$m = \frac{5}{6\lambda}$$

$$m_T = \frac{m - e^{-2\lambda T}(3/2\lambda - 2e^{-\lambda T}/3\lambda)}{1 - e^{-2\lambda T}(3 - 2e^{-\lambda T})}$$

$$\lambda_{\text{avg}}(T) = \frac{1}{m_T}$$

Similar equations can be derived for any other redundant systems. From the above examples it can be seen that at $t = 0$ the instantaneous failure rate of a redundant system is always zero. In operation this condition persists as long as at least two redundant units, each of which is capable of performing the required function alone, are operative. Also, it can be seen that the mean time between failures with preventive maintenance m_T approaches the system mean time between failures m when failed redundant components are not replaced.

The formulas given in the three examples are based on exponential components which have constant failure rates. To keep a system in this condition, besides checking and maintaining the system every T hours, we must also replace unfailed components from time to time, say every T_w hours. In practice T_w is a multiple of the inspection time T.

But the basic equations (20.7) and (20.9) for m_T and λ_{avg} are valid regardless of the failure distribution of components. For instance, if a component is known to fail only because of wearout and if it is not preventively replaced, it will fail with a mean time between failures equal to its mean wearout life M. Its instantaneous failure rate is then given by the curve in Figure 6.5, Chapter 6. However, when the component is preventively replaced every T_w hours, its mean time between failures becomes

$$m_{T_w} = \frac{\int_0^{T_w} R_w(T)\, dT}{Q_w(T_w)} \qquad (20.14)$$

where $R_w(T)$ is given by Equation (7.1b) in Chapter 7, T is the age of the component, and T_w is the regular replacement time. The average failure rate of the component is then

$$\lambda_{avg}(T_w) = \frac{1}{m_{T_w}}$$

and this failure rate can be combined in series with the failure rates of other components. Further, we can replace $R_w(T)$ by the life function $L(T)$ of a component as given by Equation (8.22) in Chapter 8, and $Q_w(T_w)$ by $1 - L(T_w)$; we then obtain from (20.14) the most general expression for the mean time between failures m_{T_w} of a component regularly replaced at T_w.

Chapter 21

COMPONENT RELIABILITY
MEASUREMENTS

B Y ITS DEFINITION, reliability is the probability of a device's performing its purpose adequately for the period of time intended under the
operating conditions encountered. Thus it is the probability of the occurrence of an event, i.e., that under certain operating conditions the device
will perform satisfactorily for a period of time. This period of time during which the device performs satisfactorily is of primary interest in reliability measurements because it is a measure of the reliability of the device.
Therefore, reliability measurements are basically time measurements from
which the probability of an event in the time domain is estimated. The
estimation falls into the category of statistical data evaluation.

To find the probability of an event, statistically significant data of
the event's occurrence must first be compiled. In the case of reliability
measurements, we have to compile statistical data on the failure-free
performance of devices in the time domain, which we do by observing
devices in operation, measuring the time of their failure-free performance,
and counting the number of failures if such should occur during the period
of observation. The time to failure t_i, which is the time of failure-free
operation from zero time t_0 until failure occurs, is an important parameter
in reliability measurements. When sufficient data about the times to
failure are available, the mean time to failure or mean time between
failures m can be closely estimated.

The term "mean time to failure" is used in the case of simple components which are not repaired when they fail but are replaced by good
specimens. If a large sample of n equal components is placed into operation at time t_0, each of these components will have a different operating

time to failure t_i. These times can be measured exactly, and the mean time to failure is then the sum of the t_i times of all n components divided by n, the number of all components. The term "mean time between failures" is used with repairable equipment or systems. Here the time t_i is the measured operating time between two failures of the equipment; it will be different for each two subsequent failures. After n failures, n measurements of the times t_i are available, and the mean time between failures m of the equipment is then the sum of the t_i times divided by the number of failures n. It is obvious that in the case of repairable equipment there is no need to measure each time between two failures separately because the sum of the times t_i is simply the total operating time T of the equipment during which the n failures occur. Thus we would simply measure the equipment's total operating time T, beginning from t_0 up to the occurrence of the nth failure, and the mean time between failures is then T/n.

It has become customary to use the term "mean time between failures" for both nonrepairable components and repairable equipment and systems. In both cases it represents the same statistical concept of the mean time at which failures occur, and the knowledge of this mean is necessary for the probability calculations which apply to the evaluation of reliability of components and systems.

We have seen in previous chapters that two types of failure may occur when components or systems built from components are in operation— chance failures and wearout failures. There is also a third type of failure, the early failure, but once these have been weeded out and if good repair policies are adopted, early failures normally should not occur in the later life of equipment. Therefore, the reliability of mature equipment is governed by the probability at which chance and wearout failures may occur. We have also seen that chance and wearout failures have different distributions. Whereas chance failures are exponentially distributed with constant failure and constant replacement rates, wearout failures are distributed normally or log normally, with a steep increase of the failure rate in the wearout period, and adopt a constant replacement rate after a so-called "stabilization period."

Because the distribution of failures in the time domain is identical to the distribution of the times to failure t_i, the times to chance failure will be distributed exponentially whereas the times to wearout failure will be distributed normally or log normally. Because the causes of these failures are different, the mean times at which these failures occur will also be different for one and the same population of components. For instance, a component population may have a mean time between chance failures m of one million hours, whereas its mean time between wearout failures may be only ten thousand hours. In this book we call the mean time between wearout failures the *mean wearout life M* of components. We use the term "life" for wearout failures, because it is wearout that definitely limits the operating life of components. Only when the chance failure rate is extremely high may chance failures override the occurrence of wearout failures, but this is usually not the case as far as the possible life of components is concerned. However, in nonrecoverable equipment built to operate only for a short time during which the probability of wearout failures can be neglected, component chance failures are the prime factor controlling the reliability of this equipment. Similarly, the reliability of equipment in which components are not allowed to wear out but are preventively replaced at regular intervals will also be governed by component chance failures.

In reliability engineering, both chance and wearout failures are of great importance. The frequency of their occurrence in the equipment jointly determines the equipment's reliability at various stages of the component's age in the equipment, but to perform the reliability calculations or to determine optimum parts replacement schedules for preventive maintenance, the distributions of the chance and wearout failures must be known separately. In other words, we should know as closely as possible the true value of the mean time between chance failures m, as well as the true values of the mean wearout life M and the standard deviation σ of the wearout failures. The parameters M and σ enable us to determine suitable replacement or overhaul schedules, and the parameter m is then used to calculate the probability that no chance failures will occur in the period between replacements or overhauls. If replacements are made only when components fail (i.e., when only repair maintenance is contemplated and no preventive maintenance is scheduled), the reliability of a long-life equipment is then determined by the parameters M and m jointly; and because M is usually much shorter than m, the reliability of such equipment stabilizes to a value governed almost exclusively by the occurrence of some kind of wearout failures of the components.

Obviously, all three parameters—m, M, and σ—of a component population change with the stress level at which the components are operated. These parameters are measured in time units, usually hours. However, the time units can also be converted into other units characteristic of the operating time and life of components, such as the number of cycles of operation (as, for instance, in the case of devices performing switching operations) or the number of revolutions (in rotating devices and ball bearings), etc. The corresponding parameters are then the mean number of cycles or revolutions between chance failures, or the mean wearout life expressed in number of operating cycles or in number of revolutions. For a known application at a given frequency of switchings or speed of rotation, these parameters are easily converted into time units.

The exact laws according to which the reliability parameters change with changing stress levels are not known as yet. In Chapter 15 we have given some empirical formulas which, with caution, can be used to good advantage. However, when high precision is required in reliability measurements, it is necessary to operate the components at the same stress levels at which they will operate in actual service. We call this procedure *laboratory reliability measurements at simulated stress conditions.* We try to simulate the entire stress spectrum as faithfully as possible for both environmental and internal stresses and measure the operating times to failure. Another method of reliability measurement is to observe components in actual service use of the equipment in which they are installed. This is a *measurement at actual service stress levels.* When the operating time is exactly measured for extended periods of operation and all failures are carefully noted, we can obtain information about the reliability of the components in the equipment and about the equipment itself. However, this is a *post factum* approach; its value lies only in the gathering of historical data or in the correcting of unreliability situations which could have been avoided at the design stage or by earlier testing of components and prototypes. Such belated corrections also involve redesign, and the procedure is expensive and time-consuming. On the other hand, historical data is very handy when no other information exists about the reliability of components. When used with judgment and caution, such data will help in the preparation of preliminary and comparative reliability analyses in new design work.

We shall now discuss various methods of reliability measurements. For this purpose we must differentiate between the probability of chance failures and the probability of wearout failures. These two probabilities provide us with two different types of information, both of which are useful for the reliable application of components. The first probability tells us how reliable components are in their useful life period, the second tells us how long components can be safely used without jeopardizing the reliability of the equipment in which they are installed. The methods of measurement and their statistical evaluation are different for chance failures and for wearout failures.

As to chance failures, we are interested in a single parameter— the mean time between failures. When this parameter is known for a given stress level of operation, the reliability at that stress level for a given mission time t is then calculated from the exponential formula $R = \exp(-t/m)$, where m, the mean time between failures, is the reciprocal of the failure rate λ. We said in Chapter 1 that the true value of a probability is theoretically never exactly known, but we can come quite close to it when we perform a large number of experiments. This also applies to the mean time between failures; therefore all we expect to obtain in reliability measurements is a reasonably good estimate. We shall use the symbol $\hat{m}$ for estimates of the true mean time between failures m.

How good an estimate is depends on the amount of available data from which the estimate is computed. We shall see later that we can set so-called "confidence limits" on both sides of the estimate, an upper and a lower confidence limit, but first let us see how to obtain estimates of the mean time between failures.

We have said that an estimate of the mean time between failures is obtained by measuring the times to failure t_i of a number of specimens, forming the sum t_i, and dividing this by the number of observations, i.e., by the number of the times to failure. However, in practice we have to bear in mind that components may fail both because of chance and because of wearout. Assume that we have a large sample of, say, 100 components from which defective and weak specimens causing early failures have been eliminated, so that we have a debugged or burned-in lot, and we have to measure the mean time between chance failures m of this lot. The main problem which we encounter right from the beginning when planning this test is how much time we can afford to spend. It is obvious that we cannot wait until all 100 components fail so as to have 100 measurements of times to failure from which to form the arithmetic mean

$$\hat{m} = \frac{\Sigma\, t_i}{n} \qquad (21.1)$$

where $n = 100$. Even if we had several years' time so that we could compute the mean for all 100 components, the question of how many of them had failed because of chance and how many had failed because of wearout would arise. We can safely assume that the majority would fail of wearout. Therefore, the mean which we obtained from Equation (21.1) would be a mean which included both chance and wearout failures and would not provide us with the required information.

We thus have to limit the duration of the test so as to be reasonably certain that no wearout failures will occur during the test period. This means that we can allow only a small fraction of the 100 components to fail, say 5, and after the fifth failure we discontinue the test. We then have 5 measurements of times to failure for all 100 components; assuming that the 5 failures were chance failures, we can compute the mean time between (chance) failures for this component population. The best estimate, called the *maximum likelihood estimate*, is obtained by adding up the operating times of all 100 components and dividing this by the number of chance failures, 5, which occurred during the test. In the numerator we have the sum total of the observed operating times of all 100 components which amounts to

$$T = t_1 + t_2 + t_3 + t_4 + t_5 + 95t_5$$

where t_1, t_2, etc., are the times at which the 5 components failed, i.e., the times during which the 5 components were in operation until they failed, and the test was discontinued at the time t_5 at which 95 components were still operating so that they accumulated an operating time of $95 \times t_5$ hours. Thus the time T is the total accumulated operating time of the 100 components during the test.

Obviously, if we were to sum up only the operating times to failure of the 5 failed components, we would get a very pessimistic result, a gross underestimate of the mean time between failures, because we neglected the fact that 95 components were still operating perfectly at the time of the test's truncation. It would be equally wrong to divide the total accumulated operating time T by the total number of components tested, because this would again ignore the fact that only 5 of them failed, not all 100 components.

If a careful inspection of the failed components revealed that one of them had failed of wearout, we would deal with this component as having been withdrawn from the test at the time of its wearout failure, but we would not count this failure at all and the total number of observed failures would reduce to 4. The total observed or accumulated time T, however, would remain the same. This procedure of accounting for withdrawals of specimens from tests, which can also occur for reasons other than wearout failures, is called *censorship*. It can be proved that this procedure is consistent with the requirement of obtaining the best estimates of the mean time between failures for the purpose of determining the chance failure rates of components.

The above method of measuring the mean time between failures of components is called the *nonreplacement* method. The failed components are not replaced during the test so that their number gradually decreases by one with each failure. Epstein has shown that when n components are originally placed under test and r of them fail at times t_1, t_2, ..., t_r counted from the beginning of the test, and the test is discontinued at the time t_r of the occurrence of the rth failure, so that $n - r$ components are still unfailed at the end of the test, the optimum estimate for the mean time between failures is given by

$$\hat{m} = \frac{t_1 + t_2 + \cdots + t_r + (n - r)t_r}{r}$$

$$= \frac{\sum_{i=1}^{r} t_i + (n - r)t_r}{r} \tag{21.2}$$

where the numerator is the total accumulated operating time T of the components under test.

If, during the test, censorship is involved on k components, because of wearout failures or other causes, so that only $r - k$ chance failures are accounted for, Equation (21.2) changes to

$$\hat{m} = \frac{\sum\limits_{i=1}^{r} t_i + (n - r)t_r}{r - k} \qquad (21.3)$$

The sum $\Sigma\, t_i$ is the operating time accumulated by the failed and withdrawn, or censored, components and r is the sum of the failed and withdrawn components. The time t_r is the time at which the test is being discontinued when the rth component fails or is withdrawn. The numerator is again the total operating time accumulated by all n components during the test.

To avoid component wearout failures during a test, the test truncation time t_r should be chosen as short as possible compared to the wearout time of the components. Wearout failures are not always easy to identify and therefore are a nuisance in measurements of the chance failure rate λ which is the reciprocal of the mean time between failures m. On the other hand, because the precision of the estimate $\hat{m}$, i.e., its nondeviation from the true value m, depends on the number of the times to failure measured during the test and therefore on the number of chance failures, it follows that the largest possible samples of components should be tested.

The choice of the sample size, i.e., of the number of components which we should submit to a test, depends on the available test time t_r and on the precision of or confidence in the test result which we wish to achieve. Suppose we expect the chance failure rate of a given type of resistor to be somewhere around $\lambda = 0.00001$, or its m to be about 100,000 hours, and we would like to get some 25 readings of times to failure so as to have a reasonably high confidence in the estimate $\hat{m}$ which we obtain after the test by the use of Equations (21.2) or (21.3). How many resistors should we use in the test if the total test time t_r at our disposal is 1000 hours? Assuming that the lot of new resistors will be defect-free and debugged, and that in the 1000 hours no resistor will fail of wearout, we want a probability of failure of $25/n$ to result from the 1000-hour test, where n is the number of components chosen to be placed initially under test. We thus expect an unreliability of

$$1 - \exp\,(-\lambda t_r) = \frac{25}{n}$$

and therefore, with $\lambda = 0.00001$ and $t_r = 1000$ hours,

$$n = \frac{25}{1 - \exp(-\lambda t_r)} = \frac{25}{1 - \exp(-0.01)} = \frac{25}{0.00995} = 2520$$

which means we would have to get about 2500 components for this 1000-hour test.

In general, when the available test time for a nonreplacement test is t hours and the expected failure rate of the specimens is λ, and m has to be measured with a precision corresponding to r chance failures, the number of specimens n to be submitted to the test is

$$n = \frac{r}{1 - \exp(-\lambda t)} = \frac{r}{Q(t)} \tag{21.4}$$

where $Q(t)$ is the expected unreliability of the components for a test operating time t.

If no provisions are made for an exact measurement of the times of each chance failure during a nonreplacement test and the test is truncated at the time t_r when the rth failure occurs in an initial lot of n components, the mean time between failures can be estimated from the formula for the probability of failure

$$\frac{r}{n} = Q(t_r) = 1 - \exp\left(-\frac{t_r}{m}\right)$$

Since the time t_r of the test duration is known and r chance failures have been counted during the test, the estimate $\hat{m}$ is obtained as

$$\hat{m} = \frac{t_r}{\ln(n) - \ln(n - r)} \tag{21.5}$$

The corresponding estimate of the per-hour failure rate is then

$$\lambda = \frac{\ln(n) - \ln(n - r)}{t_r} \tag{21.6}$$

The ratio of the failing components r to the total initial number of components n can also be expressed in terms of the percentage of failed components during the test, a, so that $a = 100r/n$. Equation (21.6) then assumes the form:

$$\lambda = \frac{\ln(100) - \ln(100 - a)}{t_r} = \frac{4.60517 - \ln(100 - a)}{t_r} \tag{21.7}$$

where a is the per cent of failed components. This formula may be handier to use because only one natural logarithm has to be evaluated.

To avoid the use of logarithms in quick estimating work, the following thumb-rule approximation of Equation (21.7) is sometimes used:

$$\lambda = \frac{a}{(100 - a/2)t} \tag{21.8}$$

where a is the per cent of components failing in a test of duration of t hours. This formula is not quite exact because it assumes that the $r = na/100$ failures occurred at an average time of $t/2$, i.e., halfway through the test, but it can be used for quick estimating work when the per cent of failing components a is small. This type of nonreplacement test, where only the total test time t is measured and the per cent of components failed is counted at the end of the test, is called the *per cent survival* or *per cent failing test method*. It has the advantage that the times of occurrence of the individual failures do not need to be measured and failed components are not replaced. Some parts manufacturers quote reliability data of components based on this per cent survival method in that they guarantee that the number of failing components will not exceed a certain percentage when the components are operated for a specified time t under specified environmental and operating conditions. This amounts to a guarantee that the failure rate will not exceed the value which can be calculated by the use of Equations (21.7) or (21.8).

A third method of estimating component reliability by measurements is that of replacement. Components which fail during the test are immediately replaced by new ones of the same population. Therefore, when n components are placed under test, the total number of components in the test remains n all the time. If the test is discontinued at the time t measured from the beginning of the test and when the rth failure has occurred, the total operating time accumulated by n components is $T = nt$ and the estimate for the mean time between failures is then*

$$\hat{m} = \frac{nt}{r} \tag{21.9}$$

where n is the number of components maintained at a constant population throughout the test.

The replacement method is seldom used in laboratory measurements because it requires constant attendance for immediate replacement of the failed components and is obviously costly. Nonreplacement tests, especially the per cent survival method, do not require constant attendance and can be run around the clock day and night. They are stopped after a predetermined test time t when the number of failed components is counted.

---

* See B. Epstein, "Truncated life tests in the exponential case," *Annals of Mathematical Statistics*, 25.3, September 1954, pp. 555–564.

However, Equation (21.9) for the replacement method is used for the evaluation of component reliability in field measurements when data on the performance of components in operating systems are being collected and processed. Assume that 100 equal components are used in a large electronic system which, since it was first put into service, has accumulated 5000 operating hours. During that time five failures of these components have occurred and have been reported. If the components are series components in the system, the total operating time accumulated by the 100 components is $100 \times 5000 = 500,000$ hours. (This is similar to a replacement test.) Therefore, the estimated mean time between failures of these components is $\hat{m} = 500,000/5 = 100,000$ hours, which corresponds to a failure rate estimate of 0.00001 per hour. Of course, confidence in this estimate will not be overwhelming, because we only have five failures to go by, but we do get a reasonably good idea about the probable reliability of these components. Or, assume that there are four magnetic amplifiers used in a temperature control system in an airplane, and 100 of these airplanes have accumulated a flight time of 300,000 flight hours since placed into service. Only one failure of that magnetic amplifier was reported during that time. The total operating time accumulated by magnetic amplifiers is thus $4 \times 300,000 = 1,200,000$ hours because there are four of them on each airplane. The estimated mean time between failures of the magnetic amplifier is then $\hat{m} = 1,200,000/1 = 1,200,000$ hours. Confidence in this result, of course, is even much lower than in the preceding example because only one failure has occurred and therefore only one reading of a time to failure is available so that no arithmetic mean can even be formed. However, the result was obtained from 401 amplifiers (the one being the replacement), each of which on the average has operated almost 3000 hours with only one failure occurring in altogether 1,200,000 hours of accumulated operating time, and the chance failure rate will therefore very probably be of the order of magnitude of 0.000001 per hour or better. In the next chapter we shall see how to assign confidence limits to estimates according to the number of failures from which the estimates were derived.

A replacement test generates information somewhat faster than a nonreplacement test, because in the former the number of components in the test is being kept constant whereas in the latter it decays exponentially. Therefore, over a period of time, more failures will occur in the replacement test. Sometimes it is asked how much time can be saved in laboratory measurements when using the more costly replacement method. Returning to the example of 2520 resistors with an expected failure rate of 0.00001 and the requirement of testing up to 25 failures, we have seen that a nonreplacement test would last about 1000 hours. A replacement test up to 25 failures would require a test time of about 990 hours, as can be ascertained from Equation (21.9). Therefore, the time saved for a large sample size is negligible, and because large sample sizes are necessary in component chance failure rate measurements in order to achieve precision and to keep the test time well below the wearout time, the obvious choice for component reliability measurements is the nonreplacement method of the per cent surviving type, unless other reasons dictate the adoption of the replacement method.

Different procedures apply to longevity measurements, i.e., measuring the mean wearout life of components and the standard deviation of wearout failures. Information about these parameters is necessary because wearout failures affect reliability in prolonged operations, and their distribution should be known also for the determination of preventive replacement or overhaul schedules. Further, it was shown in Figures 7.1 and 7.3 how the probability of wearout overrides the probability of chance failures with the age of components, and that in large complex systems with tens of thousands of components in series, even if the chance failure rates of all these components were zero, and if the systems were intended to operate only once in their lifetime for a few hours (as do missiles), the probability of wearout failure may very well become the factor dominating the unreliability of such systems. Figure 7.4 showed the probability of surviving wearout of a system consisting of 1000 components. Even if the mean wearout life of the components in the system were much longer than that shown in the figure, with ten thousand or one hundred thousand components, the curve of the system probability of surviving wearout would shift dangerously closer to the zero time point than the corresponding curve in Figure 7.4. Thus, there is no doubt that in reliability engineering the probability of component wearout failures must be very closely explored.

Measurements of the mean wearout life of components can be compared with a project to find the mean life of people dying of the effects of age. We would take a very large population sample from death statistics and would eliminate cases of premature deaths caused by accidents, epidemics, etc.; we would also eliminate cases of infantile death. A graph of the distribution of human lives, regardless of the cause of death and age, would have about the idealized form shown in Figure 21.1. In this graph the effects of aging on the frequency of deaths begin to enter the picture from the age of about 35 years. From here on "wearout failures," represented by the dashed line, are superimposed on other causes of death which gradually give place to the predominating position of wearout. We could thus assume the "wearout" deaths to be normally distributed about a mean of some 70 years with a standard deviation of about 11 years.

This example, of course, is not a perfect analogy for infantile, chance, and wearout failures of components, because the variability of humans is much more pronounced than in components of a given type. Also, when studying the failure frequency distribution of components, we are interested in finding the statistics for a given operating stress level, and we assume that the stress spectrum remains essentially constant throughout the operating life of a component. However, the example gives a good idea as to what we are after when we want to measure the mean wearout lives of components and the standard deviation of wearout lives.

To find the mean and the standard deviation of wearout failures, it is essential to start a test with new components, or, if so specified, with components which have passed a burn-in procedure of a known number

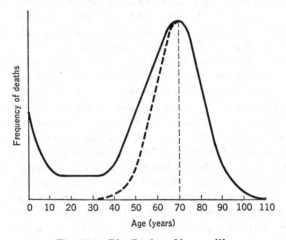

Fig. 21.1. Distribution of human life.

of operating hours to weed out early failures. Normally this would not exceed 200 hours, and thus would detract very little from component life. Furthermore, when burn-in is specified we know that the components will enter service with a certain number of burn-in hours already accumulated.

A sample of such new or burned-in components is then submitted to test operation under simulated environmental conditions, and the test continues until either all components or at least a substantial percentage of them fail. The life lived by each component is measured separately. From these readings the mean wearout life and the standard deviation are computed. If component chance failures occur during the test, such components are completely eliminated from the test results and their operating time is disregarded, in contrast to measurements of the mean time between chance failures where we include in the computations the time accumulated by withdrawn or censored components. Also, in chance failure rate measurements there is no requirement that the components be new or that the previously accumulated operating time be known, as long as we are reasonably sure that the tested components will not run into a wearout condition during the test. We assume that the chance failure rate remains constant during the useful life, and therefore the probability of chance failure for a given time does not change. But the probability of wearout failure increases with component operating age, and the total operating time lived up to wearout failure must therefore be measured.

In wearout measurements, much smaller samples will suffice than those used in chance failure rate measurements, but the test time will necessarily be much longer. As an example, let us consider a test sample of ten components of a known mean time between chance failures of $m = 200,000$ hours under certain operating conditions, and let us assume that the mean wearout life of these components is $M = 10,000$ hours under the same conditions. Thus we would expect to accumulate in the test on all ten components together a life of $10 \times 10,000 = 100,000$ operating hours if we test all of them to failure. How many of the ten components would we expect to fail because of chance before they fail of wearout? The answer is that in 100,000 accumulated operating hours we would expect not more than one chance failure to occur—and possibly none. If there were two or more chance failures in 100,000 cumulative hours, the question of whether the mean time between chance failures of such components is really 200,000 hours would necessarily arise. Of course, with 25 of these components in a life test, we would expect to accumulate together about $25 \times 10,000 = 250,000$ cumulative operating hours, and therefore we would expect more chance failures to occur— about one or two—so that we would get only 23 or 24 readings of wearout lives.

As to the expected length of a life test on ten components, whose mean wearout life is 10,000 hours, this depends on the standard deviation of lives. Assume that $\sigma = 1000$ hours. We would then very probably be through with the test in about 13,000 hours, by which time we would expect all ten components to have failed because of wearout. With a larger sample of, say, 25 components, the test time would be only somewhat longer. With only one or two specimens surviving the plus-three standard deviations time, the test would be truncated anyway.*

Further, with $\sigma = 1000$ hours we would not expect wearout failures to occur prior to 6000 or 7000 hours. Therefore, components failing earlier

---

* For the estimation of the mean and standard deviation of truncated normal distributions, see A. Hald, *Statistical Theory with Engineering Applications*, John Wiley & Sons, Inc., New York, 1952, chapter 6, section 6.9, "The Truncated Normal Distribution."

could be eliminated from the final test data as chance failures, even without exacting investigation.

For the computation of the mean wearout life and standard deviation, we use the lives of those components that we are reasonably sure failed of wearout. Physical inspection as well as careful statistical analysis of the life data, such as skewness or assymetry of the distribution, will help to eliminate chance failures if present. The equations for estimates of M and σ in hours are then

$$\hat{M} = \frac{\sum\limits_{i=1}^{r_w} t_{iw}}{r_w} \tag{21.10}$$

$$\hat{\sigma} = \sqrt{\frac{\sum\limits_{i=1}^{r_w} (t_{iw} - M)^2}{r_w}}^* \tag{21.11}$$

where t_{iw} is the operating life to wearout failure in hours of the ith component, and r_w is the number of components failed of wearout when n components were originally in test of which $n - r_w$ were chance failures. As we said before, the mean time between chance failures m of components is normally much longer than their mean wearout life M; the percentage of expected chance failures in life tests will normally be small, on the average $100M/m$ per cent, and the actual number of chance failures will therefore average out to nM/m.

When the test on a normal population is truncated so that out of n original components a components are still alive at the truncation time t_0 and we therefore have $n - a$ measurements of times to failure t_i, we obtain an estimate s of the standard deviation σ and an estimate $\hat{M}$ of the mean wearout life M as follows:

$$s = \frac{\sum\limits_{i=1}^{n-a} (t_0 - t_i)}{n - a} \cdot g(z) \tag{21.12}$$

$$\hat{M} = t_0 + zs \tag{21.13}$$

The function z, which in most tests will have a negative tabular value, is found from tables† after first computing an estimate of the degree of

---

* In statistical work, conventionally the letter s is used instead of $\hat{\sigma}$ to denote estimates of the unknown true values of the standard deviation σ. Also, in the denominator of Equation (21.11) r_w can be replaced by $r_w - 1$ to obtain the "most unbiased" estimate, or by $r_w + 1$ to obtain the "optimum" estimate of the standard deviation.

† A. Hald, *Statistical Tables and Formulas*, John Wiley & Sons, Inc., 1952, Table X, "The One-Sided Censored Normal Distribution," pp. 64 and 65. See also A. Hald, *Statistical Theory with Engineering Applications*, pp. 149 and 150.

truncation:

$$h = \frac{a}{n} \tag{21.14}$$

and

$$y = \frac{(n-a) \sum\limits_{i=1}^{n-a} (t_0 - t_i)^2}{2[\sum\limits_{i=1}^{n-a} (t_0 - t_i)]^2} \tag{21.15}$$

From the tables z is found by extrapolation when h and y are known. The value of the function $g(z)$ in Equation (21.12) is then

$$g(z) = \frac{n-a}{a\psi'(z) - (n-a)z} \tag{21.16}$$

where $\psi'(z)$ is found in the same tables by extrapolation when z is known. To reduce the variance of M and s it is necessary to continue testing until at least 70 per cent of the components have failed.

Similar to the case of the mean time between failures, the measured values of the mean wearout life M and of the standard deviation of wearout failures apply only to the stress level which existed during the measurements. As to the effects which changing stress levels have on the various reliability parameters, the reader is referred to Chapter 15, where some very general rules are given. However, it must be emphasized that the exact laws according to which reliability parameters change with changing stress levels are not known, and it is the statistical measurements at the various stress levels which can supply the best estimates. From a number of such measurements approximate formulas can be derived which allow us to estimate the parameters M and σ of a given population of components at other than measured stress levels. When the stress level is reduced, the mean wearout life becomes longer and its variance increases; when the stress level is increased, the mean wearout life is shortened and the variance becomes smaller. Referring to Figure 15.1, at a lower stress level the strength of component S_1 would start deteriorating later and with a milder slope. Derated operation therefore postpones strength deterioration and the occurrence of wearout failure.

A form of wearout is also the so-called "shelf life" of certain types of components. Components which are sensitive to environmental stresses such as humidity, normal atmospheric pressure, radiation, chemical content and impurities of the atmosphere, microbes, and air temperature may display the phenomenon of strength deterioration even if they are stored and not operating. These components, when stored for long periods of time without sufficient protection, may enter active service with their strength very much reduced owing to the cumulative damage which they suffered and can, at a given operating stress level, fail instantaneously or very early. After prolonged storage the performance parameters of such components should be checked before they are used. Statistical measurements of these parameters in the time domain on stored samples will yield estimates of the mean shelf life and of the standard deviation. In most cases normal distributions will be encountered. A useful method is to determine experimentally which of the existing stresses are most harmful to the storage of the components and to protect the components against these. Packaging in inert gases and refrigeration seem to be most promising for this purpose.

Another form of wearout is the parameter drift, which causes electronic components to drift out of tolerance, and which is a function of the operating time and stress level. In this category belong the change of capacitance of electrolytic capacitors, the change of resistance of composition resistors, the change of collector leakage current of transistors, etc. These gradual changes in themselves do not constitute component failure, but when a circuit is designed for certain tolerance limits and a component exceeds that limit, the circuit may malfunction because of that component. Thus, a system failure develops without the culprit component failing itself.

Obviously, if the circuit were designed to allow for a larger change in the component value, no circuit failure would occur or would occur only much later. Thus, in the case of component drift or tolerance degradation, we cannot strictly speak of a "component-failure" frequency. On the other hand, the degradation pattern of the tolerance limits of these components as a function of time, and also as a function of the stress level, is of the greatest importance in electronic circuit design work. This information should be supplied by the component manufacturers.

To obtain the degradation pattern a large sample of new, debugged components is placed into test operation at a specified stress level. The test environmental temperature and the electrical stresses, especially, must be specified. At the beginning of the test the characteristic parameter of all components is measured as, for instance, the collector leakage current I_{CBO} of transistors, and its distribution is plotted. After every 100, 500, or 1000 hours of test, these measurements are repeated on all components until a definite time pattern of the distribution is obtained; the pattern shows how the tolerance limits are spreading with time. The spread may be one-sided or two-sided according to the kind of components involved.

Figure 21.2 shows the generalized, one-sided degradation pattern of the collector leakage current I_{CBO} of transistors. The upper limit (100th percentile) of the leakage current of a transistor population increases linearly with time. The original distribution of the leakage current in the population is shown at the left of the graph at time $t = 0$. It can be imagined as projected into the graph from a vertical plane. In the right half of the graph the distribution is again shown after the sample has been in operation for over five time units. The upper limit of the leakage current has more than doubled. Thus, if a circuit were designed so that it allowed only for twice the original upper limit of the transistor leakage current (see the circuit design limit line), the first circuit failure due to transistor leakage current could occur at the time shown in the graph (i.e., first degradation failure). With only a single transistor of this population in the circuit, the probability of circuit failure at this time point

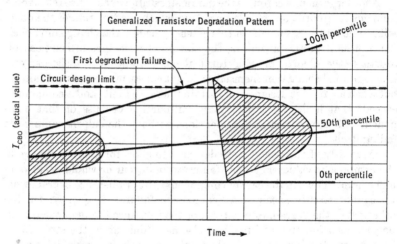

Fig. 21.2. Transistor degradation. (By courtesy of General Electric Company.)

would be quite small, depending on the extreme left tail of the I_{CBO} distribution. However, this probability increases faster as larger and larger portions of the distribution curve intersect the circuit design limit line, and for a single component it becomes 0.5 by the time when half of the area under the distribution curve has crossed the circuit design line. Thus it follows that in reliability design work, the time to the first degradation failure must be kept well above the design life of the equipment in which the circuit is to be used if degradation failures are to be avoided, or preventive replacement must be scheduled.

The graph also shows that if the circuit design limit were raised to three times the original upper limit of I_{CBO}, the time to the first degradation failure would be about doubled. Experienced designers take five times the upper limit as the circuit design limit, especially with systems which are designed to operate for a longer time, i.e., not one-shot systems.

The effect of the operating temperature on the upper tolerance limit is also of great importance. The thumb rule frequently used is that the upper limit of I_{CBO} doubles itself for every 10°C increase above the rated value.

This discussion shows the importance of knowledge of the spread vs. time or spread vs. temperature-time characteristic of component drifts for the design of reliable circuits. Graphs with actual time scales giving the test stress levels should be made available to designers by the component manufacturers' test laboratories. The users of the components have no means of performing degradation pattern tests on thousands of various components. Such graphs also allow optimum parts replacement schedules to be determined when preventive maintenance is programmed.

An analogy between ordinary wearout failures caused by strength deterioration and tolerance limit degradation failures can be mentioned at this point. Both are age-dependent events, and when they begin to show up the failure rate begins to increase more or less rapidly. Figure 21.3 shows the generalized failure rate curve of transistors. It is analogous to Figure 5.1, except that on the right side degradation failures take the place of wearout failures. The difference between wearout and degradation failures is that at a given stress level the frequency of wearout failures is strictly a function of component age alone, whereas the occur-

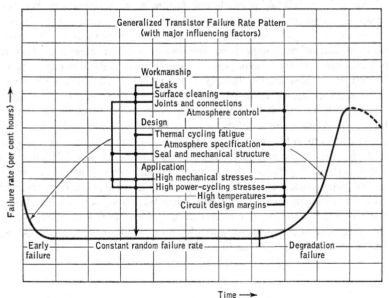

Fig. 21.3. Transistor failure rate. (By courtesy of General Electric Company.)

rence of degradation failures can be postponed by more tolerant circuit design although the stress level acting on the components remains the same. Obviously, in circuits which are not sensitive to tolerance limits, degradation failures will not occur and wearout failures will take their place, although in most cases this will happen much later.

As to stress level dependence, both types of failure begin to show up earlier when the components are operated at higher stress levels. Reduction of stress levels postpones their occurrence. Therefore, if in Figure 21.2 the stress level were increased, the 0th percentile line and the 100th percentile line would go apart faster, which means that the latter would cross the circuit design limit line earlier, and so would the 50th percentile line. The mean time to degradation failure, which can be considered to occur approximately where the 50th percentile line crosses the circuit design limit line, would thus be reduced. The distribution of degradation failures in the time domain for a given circuit design limit and operating stress level can be obtained if required from the intersection points of various percentile lines (i.e., 100th, 99th, 95th, 90th, 80th, 70th, 60th, 50th percentile) with the circuit design limit line.

Finally, we want to mention a form of comparative component reliability measurements which do not supply actual reliability estimates but give a quick answer to the question as to which of two component populations will exhibit a lower chance failure rate in operation. We take samples of two component populations—for instance, mica capacitors of the same ratings but of different manufacture—and measure the breakdown voltage of each capacitor. We then present the results in the form of histograms for each population separately and fit density curves to the histograms. These curves usually will be normal Gaussian curves with the mean breakdown voltage in the center and with a certain variance or standard deviation. By comparing the density curves of the two populations, we obtain quick information as to which brand will better withstand transient voltages and will, therefore, exhibit a lower chance failure rate. However, the actual failure rate cannot be obtained in this way, nor will this comparative strength test tell us much about the longevity of the capacitors, i.e., which of the two has a longer life to wearout. Nevertheless, this test is very useful when the choice between two brands must be made in reliability engineering and no other information about the reliability of components is available or the shortage of time does not permit regular reliability tests to be conducted.

Chapter 22

CONFIDENCE LIMITS

THE ESTIMATES OF THE MEAN TIME between failures m, the mean wearout life M, and the standard deviation s, obtained by measurements as described in the preceding chapter, are so-called *point estimates* of the true unknown parameters. How reliable are such estimates and what confidence can we have in them?

We know that statistical estimates are more likely to be close to the true value as the sample size increases. Thus, there is a close correlation between the accuracy of an estimate and the size of the sample from which it was obtained. Only an infinitely large sample size could give us a 100 per cent confidence or certainty that a measured statistical parameter coincides with the true value. In this context, "confidence" is a mathematical probability relating the mutual positions of the true value of a parameter and its estimate.

When the estimate of a parameter is obtained from a reasonably sized sample, we may logically assume that the true value of that parameter will be somewhere in the neighborhood of the estimate, to the right or to the left. Therefore, it would be more meaningful to express statistical estimates in terms of a range or interval with an associated probability or confidence that the true value lies within such interval than to express them as point estimates. This is exactly what we are doing when we assign confidence limits to point estimates obtained from statistical measurements.

Confidence intervals around point estimates have a lower confidence limit L and an upper confidence limit U. If, for instance, we calculate the confidence limits for a probability of, say, 90 per cent, this means that in 90 per cent of the cases the true value will lie within the calculated limits, whereas in 10 per cent of the cases it will lie outside these limits,

227

either below or above. The 90 per cent confidence that the true value lies
within the calculated limits is called the *confidence level.* Obviously, if we
wished to increase the confidence level to, say, 99 per cent, so that in
99 per cent of the cases the true value would lie within the confidence
limits, the confidence interval around the point estimate would become
much wider—or we would have to use a much larger sample for the point
estimate.

To calculate the confidence limits we must know more about the statistical distribution of the estimate. The concept of the statistical distribution of an estimate can be readily explained on the mean value of a
normal distribution. If we take a random sample of n components from
a large population in which the occurrence of failures is known to be
normally distributed (wearout), and we run a life test until each component in the sample fails so that n failures, and therefore n measurements
of the times to wearout failure are available, we compute the mean life
of the sample as

$$M = \Sigma \frac{t_i}{n}$$

and the standard deviation of component life as

$$\sigma = \sqrt{\Sigma \frac{(t_i - M)^2}{n}}$$

These values are the true parameters of the sample but it is very unlikely
that they are the true parameters of the large population from which the
sample was taken. M is only an estimate of the population's true mean
life, and an unbiased estimate of the population's standard deviation is
obtained as

$$s = \sqrt{\Sigma \frac{(t_i - M)^2}{n - 1}}$$

If we take several more samples of n components from the same population, each of these samples will yield more or less different values of
M and s. Obviously, the obtained estimates of M will be distributed in
some manner about the true M of the population, and the estimates of s
will be distributed in some manner about the true σ of the component life.
Intuitively we see immediately that the variance at which the mean estimates are distributed about the true mean is necessarily smaller than the
variance at which individual component lives (t_i) are distributed about M.

If the population is normal, the mean life estimates obtained from
several samples are again normally distributed about the true M of the
population.* Thus, we say that the estimated mean of a normal popu-

* This treatment of confidence limits is based on *Engineering Statistics* by A. H.
Bowker and G. J. Lieberman, Prentice-Hall, Inc., 1959, chap. 8.

lation is normally distributed about the true mean. However, whereas the life of individual components is distributed about M with a standard deviation σ, a mean life estimate obtained from a sample of n components with n failures counted is distributed about the true mean life of the large population with a standard deviation of

$$\sigma(M) = \frac{\sigma}{\sqrt{n}} \qquad (22.1)$$

where σ is the true standard deviation of component life. The standard deviation $\sigma(M)$ of the mean is called *standard error*. We use here the notation $\sigma(M)$ only for the purpose of clearly differentiating between the standard deviation of component life and that of mean estimates.

Equation (22.1) immediately allows us to assign confidence limits to an estimated mean obtained from a large sample. From the standardized normal curve we know that the true M will lie within ± 1 standard error $\sigma(M)$ of the measured estimate $\hat{M}$ in about 68.3 per cent of the cases, within $\pm 2\sigma(M)$ in about 95.4 per cent of the cases, and within $\pm 3\sigma(M)$ in about 99.7 per cent of the cases. The per cent probabilities associated with an interval $\pm K\sigma(M)$, i.e., 68.3, 95.4, 99.7 per cent, are the confidence levels for the respective intervals and correspond to the areas under the normal curve between the respective interval limits. The coefficient K is the percentage point on the abscissa of the standardized normal curve for a certain area under the curve. K is thus the number of standard deviations $\sigma(M)$ from the mean and is obtained from normal tables for any required probability. In our case, K indicates how many standard errors we have to subtract and to add to our estimate $\hat{M}$ to obtain the lower and upper confidence limits for a required confidence level. The confidence interval is then given by

$$\hat{M} \pm K\sigma(M) = \hat{M} \pm K\frac{\sigma}{\sqrt{n}} \qquad (22.2)$$

Normal tables usually give the tail areas under the normal distribution curve. For $K = 1$, i.e., one standard deviation from the mean, from tables we find the tail area as 0.1587. Because the curve has two tails, the area outside of a range of ± 1 standard deviation is $2 \times 0.1587 = 0.3174$. The area within that range is then $1 - 0.3174 = 0.6826$, or about 68.3 per cent. If we denote the area under one tail by $\alpha/2$ at the percentage point $K = K_{\alpha/2}$, the area within the interval $\pm K_{\alpha/2}\sigma(M)$ will be $(1 - \alpha)$. The corresponding confidence level is then $100(1 - \alpha)$ per cent. Figure 22.1 shows this relationship graphically. In tables the notations x, z, or u are frequently used instead of K. In Figure 22.1 when we put the mean life estimate of $\hat{M}$ hours in the point of origin O, as shown for instance in Figure 6.4, the lower confidence limit will be

$$L = \hat{M} - K_{\alpha/2}\sigma(M) = \hat{M} - K_{\alpha/2}\frac{\sigma}{\sqrt{n}} \qquad (22.3)$$

and the upper confidence limit,

$$U = \hat{M} + K_{\alpha/2}\sigma(M) = \hat{M} + K_{\alpha/2}\frac{\sigma}{\sqrt{n}} \qquad (22.4)$$

When the estimate M is in hours, $\sigma(M)$ will also be in hours, and L and U will be in hours. The range between L and U will include M with a probability $(1 - \alpha)$, which is expressed by the following probability equation:

$$P\left(\hat{M} - K_{\alpha/2}\frac{\sigma}{\sqrt{n}} \leqq M \leqq \hat{M} + K_{\alpha/2}\frac{\sigma}{\sqrt{n}}\right) = 1 - \alpha \qquad (22.5)$$

When the lives of n components are known from a wearout test and we compute their mean M and their standard deviation s, and when n is large so that we can assume that $s \approx \sigma$, the upper and lower confidence limits can be readily evaluated from the following table for any of the confidence levels listed:

$K_{\alpha/2}$	Two-sided confidence intervals $\hat{M} \pm K_{\alpha/2}s/\sqrt{n}$	Confidence levels $100(1 - \alpha)\%$
0.84	$\hat{M} \pm 0.84s/\sqrt{n}$	60.0
1.00	$\hat{M} \pm 1.00s/\sqrt{n}$	68.3
1.28	$\hat{M} \pm 1.28s/\sqrt{n}$	80.0
1.50	$\hat{M} \pm 1.50s/\sqrt{n}$	86.6
1.64	$\hat{M} \pm 1.64s/\sqrt{n}$	90.0
1.96	$\hat{M} \pm 1.96s/\sqrt{n}$	95.0
2.00	$\hat{M} \pm 2.00s/\sqrt{n}$	95.4
2.58	$\hat{M} \pm 2.58s/\sqrt{n}$	99.0
3.00	$\hat{M} \pm 3.00s/\sqrt{n}$	99.7
3.29	$\hat{M} \pm 3.29s/\sqrt{n}$	99.9

Strictly speaking, this procedure of assigning confidence intervals to an estimate is correct only when the true standard deviation σ of component wearout lives is known and used instead of s in the above table. However, it can be applied in reliability work as an approximation whenever the estimate s of σ was obtained from a large sample, i.e., when the number of failures is at least 25, and preferably, more.

Actually, in reliability work we are usually more interested in the lower confidence limit L of the mean wearout life than in the upper limit. When L is computed as shown above, we subtract from it that number k of component life standard deviations σ which corresponds to a given reliability requirement (see Chapters 6, 7, 8, and 20), and we obtain the component replacement or overhaul time as

$$T_o = L - k\sigma \qquad (22.6)$$

In a similar way we gain assurance that a nonrepairable system (for instance, a missile) will not fail because of wearout of a specific component during its mission time.

When only the lower confidence limit L is of interest, we apply the procedure of so-called "one-sided" confidence limits. In specifications we often find the requirement that the mean wearout life must exceed a

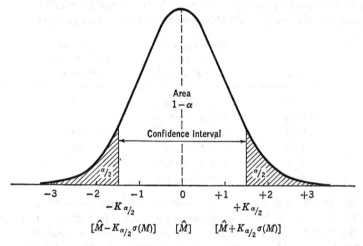

Fig. 22.1. Two-sided confidence level, interval, and limits.

specified minimum value with a stated confidence level of $100(1 - \alpha)$ per cent. We consider the specified minimum value as the lower confidence limit L and let the upper limit extend to infinity. The problem is to assure the customer that with a probability $(1 - \alpha)$, the true mean life is equal to or larger than the specified minimum. This means that there is a risk, or probability α that the true M will be less than the specified minimum.

Whereas in the case of two-sided confidence limits we had an area of $\alpha/2$ under the left tail of the normal curve, we now have an area α to the left of L and the area $(1 - \alpha)$ to the right. The percentage point corresponding to L therefore becomes K_α and we can write

$$L = \hat{M} - K_\alpha \frac{\sigma}{\sqrt{n}} \qquad (22.7)$$

Therefore the estimate of mean life obtained from a qualification test should be

$$\hat{M} \geqq L + K_\alpha \frac{\sigma}{\sqrt{n}} \tag{22.8}$$

If this equation is not satisfied, i.e., if the test result shows that $\hat{M}$ is smaller than required by the equation, the customer requirement that the true M must be at least L at the specified $100(1 - \alpha)$ per cent confidence level has not been fulfilled.

The following table, in which the assumption $s \approx \sigma$ is made, allows a quick check as to whether an estimate $\hat{M}$ obtained from a sample size n fulfills the requirement that the true M must not be smaller than the specified minimum L. Only the usual confidence levels are given.

K_α	The estimate $\hat{M}$ must exceed: $L + K_\alpha s/\sqrt{n}$	Specified confidence level $100(1 - \alpha)\%$
0.25	$L + 0.25s/\sqrt{n}$	60
0.52	$L + 0.52s/\sqrt{n}$	70
0.84	$L + 0.84s/\sqrt{n}$	80
1.28	$L + 1.28s/\sqrt{n}$	90
1.64	$L + 1.64s/\sqrt{n}$	95
2.33	$L + 2.33s/\sqrt{n}$	99

As previously mentioned, the above procedure can be applied if the sample size n is at least 25. However, similar procedures also apply to smaller sample sizes except that now we cannot assume that $s \approx \sigma$, and we must use another set of equations based on Student's t distribution. Actually, all we do is replace the normal percentage points $K_{\alpha/2}$ and K_α in the above-developed equations by the tabulated percentage points $t_{\alpha/2;n-1}$ and $t_{\alpha;n-1}$ of the t distribution, where $n - 1$ is called the *degrees of freedom* and n is the number of failures.

We obtain the following confidence limits for a two-sided confidence interval at a confidence level of $100(1 - \alpha)$ per cent:

$$L, U = \hat{M} \pm t_{\alpha/2;n-1} \frac{s}{\sqrt{n}} \tag{22.9}$$

and for a one-sided lower limit at the same confidence level:

$$L = \hat{M} - t_{\alpha;n-1} \frac{s}{\sqrt{n}} \tag{22.10}$$

Equation (22.8), which defines the permissible minimum value of the estimate $\hat{M}$ with respect to the minimum specified value L of the true mean life M, becomes:

$$\hat{M} \geq L + t_{\alpha;n-1} \frac{s}{\sqrt{n}} \tag{22.11}$$

When the problem is to determine component replacement or overhaul time from the parameters M and s obtained from only a small sample, the assumption $s \approx \sigma$ cannot be made and the component standard deviation in Equation (22.6) has to be replaced by its upper confidence limit S_u. The calculation of confidence limits for standard deviations is based on the chi-square distribution*. It can be shown that the variance ratio

$$\frac{(n-1)s^2}{\sigma^2}$$

has a chi-square distribution with $(n-1)$ degrees of freedom. The upper one-sided confidence limit for component standard deviation, for a confidence level $(1-\alpha)$, is therefore

$$S_u = s \sqrt{\frac{n-1}{\chi^2_{1-\alpha;n-1}}} \tag{22.12}$$

In this equation s is the estimate of the component standard deviation obtained from a test on n components which produced n measurements of component wearout lives t_i, i.e.,

$$s = \sqrt{\sum \frac{(t_i - M)^2}{n-1}}$$

and the percentage point $\chi^2_{1-\alpha;n-1}$ for $n-1$ degrees of freedom is found from chi-square tables.

Equation (22.12) then gives a $100(1-\alpha)$ per cent assurance that the true standard deviation of component life does not exceed the value S_u. Thus, for small samples the replacement or overhaul time equation (22.6) assumes the form:

$$T_o = L - kS_u \tag{22.13}$$

When the estimates $\hat{M}$ and s are derived from truncated tests according to Equations (21.12) and (21.13) and we have to assign confidence limits to these estimates, the procedure becomes more complicated. The problem is that of finding the standard deviation $\sigma(M)$ for the truncated mean estimate $\hat{M}$. We know that in a nontruncated sample, $\sigma(M) = \sigma/\sqrt{n}$. The simplest method in the truncated case known to the author

* Bowker and Lieberman, *Engineering Statistics*, pp. 86 and 87, and p. 219.

is that developed by Hald.* According to this we apply a correction factor $\mu_{11}(z)$,† the magnitude of which is determined by the value of z in Equation (21.13). In Equations (22.1) through (22.8) and in the tables belonging to these equations we then substitute the corrected standard deviation of the mean

$$\sigma(M) = \sqrt{\frac{s^2}{n} \cdot \mu_{11}(z)} \qquad (22.14)$$

for the value $\sigma(M) = \sigma/\sqrt{n} \approx s/\sqrt{n}$.

For instance, Equation (22.8) for the one-sided lower confidence limit becomes

$$\hat{M} \geqq L + K_\alpha \sqrt{\frac{s^2}{n} \mu_{11}(z)} \qquad (22.15)$$

where $\hat{M}$ is given by Equation (21.13) as $\hat{M} = t_0 + zs$, and s is given by Equation (21.12). Small sample tests are usually not truncated, but if so, probably the best procedure would be to apply again Hald's correction factor to Equations (22.9) through (22.11).

As to the upper confidence limit of an estimate s of the standard deviation obtained from a truncated test, we would use Equation (22.12), i.e., the same as for a nontruncated test.‡

The calculation of confidence limits is much simpler in the case of the one-parameter exponential distribution. The task is to assign confidence limits to an estimate $\hat{m}$ of the true mean time between failures, when $\hat{m}$ was obtained from a test in which r failures were counted. We use here the chi-square distribution.

It has been shown that the ratio

$$2r \frac{\hat{m}}{m}$$

has a chi-square distribution with $2r$ degrees of freedom when the test from which the estimate $\hat{m}$ was obtained was terminated as the rth failure occurred.§ For a two-sided confidence level $(1 - \alpha)$ we can write the following probability equation, making use of the $\alpha/2$ and $1 - \alpha/2$ per-

---

* A. Hald, *Statistical Theory with Engineering Applications*, John Wiley & Sons, Inc., New York, 1952, pp. 245 and 246.

† The factor $\mu_{11}(z)$ is found from Table X of A. Hald's *Statistical Tables and Formulas*, John Wiley & Sons, Inc., New York, 1952, pp. 64–65.

‡ However, for large samples a specific method is given in Hald, *Statistical Theory*, pp. 316–318.

§ B. Epstein, "Estimation From Life Test Data," *IRE Transactions on Reliability and Quality Control*, Vol. RQC-9 (April 1960).

centage points of the chi-square distributions:

$$P\left(\chi^2_{1-\alpha/2;2r} \leqq \frac{2r\hat{m}}{m} \leqq \chi^2_{\alpha/2;2r}\right) = 1 - \alpha \qquad (22.16)$$

which means there is a probability $(1 - \alpha)$ that the value of the ratio $2r\hat{m}/m$ will be within the interval given by the two percentage points. By rearrangement we can write:

$$P\left(\frac{2r\hat{m}}{\chi^2_{\alpha/2;2r}} \leqq m \leqq \frac{2r\hat{m}}{\chi^2_{1-\alpha/2;2r}}\right) = 1 - \alpha \qquad (22.17)$$

or, for a two-sided confidence interval at a confidence level of $100(1 - \alpha)$ per cent we obtain

$$\hat{m}\frac{2r}{\chi^2_{\alpha/2;2r}} \leqq m \leqq \hat{m}\frac{2r}{\chi^2_{1-\alpha/2;2r}} \qquad (22.18)$$

Then the two-sided lower confidence limit is

$$L = \frac{2r}{\chi^2_{\alpha/2;2r}}\hat{m} = \frac{2T}{\chi^2_{\alpha/2;2r}} \qquad (22.19)$$

and the upper confidence limit is

$$U = \frac{2r}{\chi^2_{1-\alpha/2;2r}}\hat{m} = \frac{2T}{\chi^2_{1-\alpha/2;2r}} \qquad (22.20)$$

The estimate $\hat{m}$ in the above formulas is defined as

$$\hat{m} = \frac{T}{r} \qquad (22.21)$$

and can be derived from either a replacement or a nonreplacement test. T is the sum of the operating times accumulated by all the components during the test, and is therefore the total observed operating time $\Sigma\, t_i$, where t_i is the measured operating time of the ith component. For a single system T is the straight operating time measured up to the occurrence of the rth failure. From (22.21) it follows that $T = \hat{m}r$, and therefore we can replace $2r\hat{m}$ by $2T$ in all the above equations, as shown in (22.19) and (22.20).*

When it is required that the true mean time between failures must exceed a specified minimum value with a probability of $(1 - \alpha)$, i.e., at a confidence level of $100(1 - \alpha)$ per cent, and the actual value of m is of no further interest except that it must lie somewhere above the specified minimum, we use the procedure of a one-sided confidence limit by applying a lower chi-square percentage which corresponds to a tail area α instead of $\alpha/2$; this means there is a probability of 100α per cent that m will be smaller than the specified minimum and a probability of

---

* Sir Ronald A. Fisher, *Statistical Methods and Scientific Inference*, Oliver & Boyd, Edinburgh-London, 1956, p. 53.

$100(1 - \alpha)$ per cent that it will be larger. We thus identify the specified minimum with the one-sided lower confidence limit, for which we use here the notation C_L to distinguish it from the two-sided confidence limit L, and obtain

$$C_L = \frac{2r}{\chi^2_{\alpha;2r}} \, \hat{m} = \frac{2T}{\chi^2_{\alpha;2r}} \tag{22.22}$$

We then must prove in a test that

$$\hat{m} \geqq C_L \frac{\chi^2_{\alpha;2r}}{2r} \tag{22.23}$$

or, that in an accumulated test time of

$$T = C_L \frac{\chi^2_{\alpha;2r}}{2} \tag{22.24}$$

not more than r failures have occurred. For the number of failures r any integer number from 1 upwards can be chosen. From chi-square tables we find the α percentage point corresponding to the specified confidence level of $100(1 - \alpha)$ per cent and to the chosen r, i.e.,

$$\chi^2_{\alpha;2r}$$

plug this value into Equation (22.24), and because C_L is the specified minimum value of the true mean time between failures, we obtain T in hours if C_L is specified in hours. The evaluated T also determines the sample size in unit-hours for the test.

For example, if we choose $r = 1$ and the specified confidence level is 0.9 or 90 per cent, we find from chi-square tables the percentage point corresponding to $\alpha = 0.1$, because $(1 - 0.1) = 0.9$, and to $2r = 2$ degrees of freedom. The tabular value is 4.605. According to Equation (22.24) we take one-half of this value, i.e., 2.3, which means that in an accumulated test time of $T = 2.3C_L$ there must be not more than one failure. The formula requires that T be accumulated in the test. Thus we could take one component or equipment and operate it without failure for $2.3C_L$ hours. Or, if the equipment is repairable and it fails after x hours of test, we repair it and continue the test for the remaining $t_r = 2.3C_L - x$ hours without failure. Or, we start the test with n components or equipments to reduce the straight test time and measure the operating time of each, until the sum of these times is $2.3C_L$ and no failure, or at the most one failure, occurs. If one of the specimens fails at x hours, we have by that time accumulated nx hours of operating time. We therefore shall continue the test for the remaining

$$t_r = \frac{2.3C_L - nx}{n - 1} \text{ hours} \tag{22.25}$$

because we are left with only $n - 1$ specimens in the test. During the time t_r no second failure must occur. The straight test time t for a successful test with n components or equipments will be

$$t = x + \frac{2.3C_L - nx}{n - 1} = \frac{2.3C_L - x}{n - 1} \text{ hours} \qquad (22.26)$$

where x is the operating time at which the one specimen failed. In the extreme that no failure occurs during the entire test the straight test time reduces to

$$t = \frac{2.3C_L}{n} \text{ hours} \qquad (22.27)$$

and if the one failure occurs immediately when the test is started, i.e., practically at time $x = 0$, the straight test time becomes

$$t = \frac{2.3C_L}{n - 1} \text{ hours} \qquad (22.28)$$

Thus, we see that with several components we can reduce the straight test time well below the specified minimum mean time between failures C_L. Actually, with a chosen $r = 1$ and the specified confidence level of 90 per cent, with four specimens in test we would already reduce the test time to $0.77C_L$ in the worst case, i.e., when one specimen fails at $x = 0$ hours, and to $0.58C_L$ if there is no failure.* As we shall soon see, if there is no failure we can terminate the test even earlier. In Chapter 24 we shall show that when we consider the so-called "consumer" and "producer" risks," we may extend the test time if two failures occur before the time T in Equation (22.24) is accumulated and still hope for a favorable outcome of the test.

For a 95 per cent confidence level the chi-square percentage point would be 5.991, of which one-half is about 3, and we would replace in the above equations the 2.3 coefficient in front of C_L in the above equations by the coefficient 3.0. Similar equations can be developed if the number of chosen failures is $r = 2, 3, 4$, etc. Increasing the number of specimens in test also helps reduce the test time drastically. Another advantage of having several specimens in test is that in this way the variability among them is considered.

Equations (22.17) through (22.20) pertaining to the two-sided confidence limits, and (22.22), which gives the lower one-sided confidence limit for m, can also be given the following reliability interpretation. We can have a $100(1 - \alpha)$ per cent confidence that the true reliability R is in

---

* If the specimens under test are subject to wearout and have a scheduled replacement or overhaul time shorter than the test time t, they should be reconditioned or replaced by new specimens at the scheduled time before the test can continue.

the interval between exp $(-t/L)$ and exp $(-t/U)$, or that it is larger than or equal to exp $(-t/C_L)$.

So far we have discussed confidence limits on m when the estimate $\hat{m}$ was obtained from a test terminated at the time of the occurrence of the rth failure. However, a test can also be terminated at some preselected test time without a failure's occurring exactly at that time. Epstein has shown that if we accumulate $T = \Sigma t_i$ hours of operating time and observe r failures in such a test, which can be of the replacement or nonreplacement type, the two-sided confidence limits for a $100(1 - \alpha)$ per cent confidence level are given by

$$\frac{2T}{\chi^2_{\alpha/2;2r+2}} \leqq m \leqq \frac{2T}{\chi^2_{1-\alpha/2;2r}} \tag{22.29}$$

and the one-sided confidence limit at the same level is given by

$$m \geqq \frac{2T}{\chi^2_{\alpha;2r+2}} \tag{22.30}$$

We see from these equations that even if no failure occurs during the test ($r = 0$), a definite lower confidence limit can be calculated. In such a case Equation (22.30) produces the lower one-sided confidence limit of

$$C_L = \frac{2T}{\chi^2_{\alpha;2}} \tag{22.31}$$

Further, there is the per cent survival method (see Chapter 21) in which the accumulated operating time T is not measured and only the straight test duration time t_d is known, at the end of which r failures out of n specimens in test are counted. Here we obtain at a one-sided confidence level of $100(1 - \alpha)$ per cent the following lower limit estimate of reliability for t_d hours:*

$$\hat{R}(t_d) = \frac{1}{1 + \left(\dfrac{r+1}{n-r}\right) F_{\alpha;2r+2;2n-2r}} \tag{22.32}$$

where F is the upper α percentage point of the F distribution, with the corresponding degrees of freedom. We can make the following statement about this estimate of reliability: There is a probability of $1 - \alpha$ that the true reliability for t_d hours is equal to or larger than $\hat{R}(t_d)$. This reliability estimate is nonparametric and valid for the exponential as well as the nonexponential case.

If the case is exponential and when we use the notation Y for the whole denominator on the right side of Equation (22.32), we obtain the

* See B. Epstein, "Estimation From Life Test Data."

$100(1 - \alpha)$ per cent one-sided lower confidence limit for the mean time between failures from

$$m \geqq \frac{t_d}{\ln Y} \qquad (22.33)$$

If no failures occurred in the test time t_d, Equation (22.32) reduces to

$$\hat{R}(t_d) = \frac{1}{1 + \dfrac{1}{n} F_{\alpha;2;2n}} \qquad (22.34)$$

and Equation (22.33) reduces to

$$m \geqq \frac{t_d}{\ln \left(1 + \dfrac{1}{n} F_{\alpha;2;2n}\right)} \qquad (22.35)$$

For large values of n we find the F percentage point $F_{\alpha;2;\infty}$.

Thus, if a per cent survival test does not result in any failures during the test duration t_d, we can again calculate the lower limit of the true mean time between failures and of the reliability $R(t_d)$ for any specified confidence level.

Equations (22.32) through (22.35) also apply to cases when a single equipment completes n missions each of a duration t_d. They also apply to several equal equipments in simultaneous or nonsimultaneous operation, which together accumulate n missions of equal length t_d each.

Finally, if confidence limits have to be assigned to reliability estimates $\hat{R}$ of one-shot equipment (like missiles), where the estimate $\hat{R}$ is obtained as a probability from $\hat{R} = (n - r)/n$ when n equipments are tested and r of them fail, so that $n - r$ complete the test successfully, we know that, as in the example of coin tossing, the probability estimate has a binomial distribution about the true R, with a standard deviation $\sigma = \sqrt{nRQ}$, where Q is the unreliability $1 - R$. We then apply the familiar methods used in quality control procedures.

The following examples show the use of the equations for confidence limits given on the preceding pages.

An exponentially failing large electronic system showed $r = 20$ failures in 2000 hours of operation. The 20th failure occurred at exactly $T = 2000$ hours when the observation was terminated. What statement can be made about the true mean time between failures of the system at a 95 per cent confidence level?

First, a point estimate is obtained:

$$m = \frac{T}{r} = \frac{2000}{20} = 100 \text{ hours}$$

Then we calculate the lower and upper confidence limits according to Equations (22.19) and (22.20) for $1 - \alpha = 0.95$, i.e., for the percentage points $\alpha/2 = 0.025$ and $1 - \alpha/2 = 0.975$, and for $2r = 40$ degrees of freedom. From chi-square tables* we find

$$\chi^2_{0.025;40} = 59.3 \quad \text{and} \quad \chi^2_{0.975;40} = 24.4$$

Then the lower and upper confidence limits are

$$L = \frac{40 \times 100}{59.3} = 67 \text{ hours} \quad \text{and} \quad U = \frac{40 \times 100}{24.4} = 164 \text{ hours}$$

Thus we can make the statement that there is a probability of 95 per cent that the true mean time between failures of the system is between 67 hours and 164 hours.

To find the two-sided confidence limits we can use the graph in Figure 22.2 instead of chi-square tables.† This graph shows the per cent deviation of the lower and upper limits from a point estimate m for several confidence levels of $100(1 - \alpha)$ per cent for up to 1000 failures observed.

In our example the lower limit of 67 hours deviates from $m = 100$ hours by -33 per cent, whereas the upper limit of 164 hours deviates from m by $+64$ per cent. These deviations can be obtained directly from the 95 per cent confidence level curves in the graph for $r = 20$ failures.

If the required confidence level were only 80 per cent, the 80 per cent curves in the graph show for $r = 20$ a minus deviation of 23 per cent and a plus deviation of 37 per cent from m. Thus, the lower 80 per cent confidence limit becomes $100 - 0.23 \times 100 = 77$ hours, and the upper 80 per cent confidence limit becomes $100 + 0.37 \times 100 = 137$ hours.

We can then make the statement that there is an 80 per cent probability that the true mean time between failures for which we obtained a point estimate of $m = 100$ hours based on 20 failures lies between 77 hours and 137 hours.

Now let us assume that a customer has originally specified that the system's mean time between failures must exceed 70 hours at a 95 per cent confidence level. This is a one-sided confidence requirement, which permits a probability of 5 per cent that m will be less than 70 hours. We have $1 - \alpha = 0.95$ and $\alpha = 0.05$ for the one-sided upper chi-square percentage point:

$$\chi^2_{0.05;40} = 55.8$$

---

* Attention is drawn to the fact that some tables give the $1 - \alpha$ percentage points; others give the α percentage points.

† Reproduced from the "Boeing Airplane Company's Reliability Handbook," Document D6-2770 (1957).

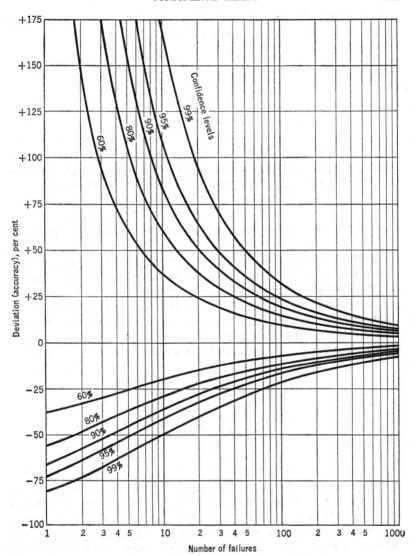

Fig. 22.2. Confidence limits for measurement of mean-time-between-failures.

According to Equation (22.22) we obtain the 95 per cent lower confidence limit as

$$C_L = \frac{40 \times 100}{55.8} = 72 \text{ hours}$$

Thus the customer's requirement is satisfied.

From Figure 22.2 we also can obtain one-sided confidence limits by the use of the following conversion table of confidence levels:

Two-sided confidence level (%)	One-sided confidence level (%)
60	80
80	90
90	95
95	97.5
99	99.5

For a lower one-sided 95 per cent requirement in our example, we would use the 90 per cent curve in lower portion of the graph which for $r = 20$ shows a deviation of -28 per cent from m, and therefore

$$C_L = m - 0.28m = 72 \text{ hours}$$

The result is the same as previously calculated by means of chi-square tables.

Let us now assume that in the above example the 20th failure occurred some time prior to $T = 2000$ hours, but we terminated our observations at 2000 hours without a 21st failure's occurring. Thus, all we know now is that we observed a total of 2000 equipment operating hours during which time 20 failures occurred. Let us now assign 95 per cent two-sided and lower one-sided confidence limits to the mean time between failures. Using Equations (22.29) and (22.31) we obtain

$$L = \frac{2T}{\chi^2_{0.025;42}} = \frac{4000}{61.8} = 65 \text{ hours}$$

$$U = \frac{2T}{\chi^2_{0.975;40}} = \frac{4000}{24.4} = 164 \text{ hours}$$

$$C_L = \frac{2T}{\chi^2_{0.05;42}} = \frac{4000}{58.1} = 69 \text{ hours}$$

We can therefore make the statements that with a probability of 95 per cent the true m of the system lies between 65 and 164 hours, or it is larger than $C_L = 69$ hours. We can further state that with a probability of 97.5 per cent the true m is larger than $L = 65$ hours. Note that in this calculation we used $2r + 2 = 42$ degrees of freedom for the lower limits.

As another example we consider the case of 400 magnetic amplifiers in an aircraft temperature control system. Four of these amplifiers are used in an airplane, and 100 of these airplanes have accumulated a total flight time of 300,000 hours. The magnetic amplifiers have therefore accumulated a total operating time of 1,200,000 hours with only one failure reported which occurred during a 5-hour flight after which the magnetic amplifier was replaced. Strictly taken, we should subtract

5 hours from the total operating time of 1,200,000 hours if the amplifier failed at the beginning of that flight. We see, however, that in this case the 5 hours are of no consequence.

Let us now find the 80 per cent confidence limits:

$$L = \frac{2,400,000}{\chi^2_{0.1;2r+2}} = \frac{2,400,000}{\chi^2_{0.1;4}} = \frac{2,400,000}{7.779} = 308,522 \text{ hours}$$

$$U = \frac{2,400,000}{\chi^2_{0.9;2r}} = \frac{2,400,000}{\chi^2_{0.9;2}} = \frac{2,400,000}{0.211} = 11,374,408 \text{ hours}$$

$$C_L = \frac{2,400,000}{\chi^2_{0.2;2r+2}} = \frac{2,400,000}{\chi^2_{0.2;4}} = \frac{2,400,000}{5.989} = 400,634 \text{ hours}$$

And, if no failure had occurred in the 1,200,000 hours:

$$L = \frac{2,400,000}{\chi^2_{0.1;2}} = \frac{2,400,000}{4.605} = 521,172 \text{ hours}$$

$$U = \frac{2,400,000}{0} = \text{infinity}$$

$$C_L = \frac{2,400,000}{\chi^2_{0.2;2}} = \frac{2,400,000}{3.219} = 745,574 \text{ hours}$$

We see in this example that when no failure has occurred, we are not in a position to calculate a point estimate m, but we still can calculate confidence limits for any required confidence level.

As a final example, let us consider a component with a known mean time between chance failures m of, say, 100,000 hours, but having a mean wearout life M of only 2000 hours with a standard deviation of 200 hours. The component is easily accessible and is replaced by plugging in a spare when it fails. How many spares must be carried if the component is part of a space ship's internal system, so as to be 99 per cent sure that a spare will always be available during an interplanetary trip of 20,000 hours' maximum duration?

Let us first assume that only wearout failures occur. According to Equation (22.2) the confidence interval after n failures is

$$\mathfrak{M} \pm K\sigma \sqrt{n}$$

where $\mathfrak{M} = nM$ is the grand total mean, or the expected total life of n components operating in succession. As the problem is stated, we want to be 99 per cent sure that the lower one-sided confidence limit

$$L = nM - K_{0.01}\sigma \sqrt{n}$$

is at least 20,000 hours.

Knowing the values of M and σ, we can calculate n, the number of components required for 20,000 hours' operation at the 99 per cent one-sided confidence level, from

$$20{,}000 = n2000 - (2.33)(200)\sqrt{n}$$

Solving this equation for the nearest larger integer of n, we obtain

$$n = 11$$

Thus, we need 11 components to achieve a total life of 20,000 hours at a confidence level of at least 99 per cent. With the first component installed in the system at the time of departure we need

$$n - 1 = 10$$

spares to take with us on the trip. But $n = 11$ gives us a considerably higher confidence, or reliability in this case, than the required 99 per cent. The actual reliability of 11 successively operating components for an operating time of 20,000 hours amounts to

$$20{,}000 = 11 \times 2000 - K_\alpha 200 \sqrt{11}$$

$$K_\alpha = \frac{2000}{(200)(3.22)} = 3.012$$

$$R = 1 - \alpha = 0.9987$$

On the other hand, with only nine spares so that $n = 10$, we would not achieve the required confidence or reliability of 0.99, but only of 0.5.

Next, let us assume that only chance failures will occur. We need x components, and therefore $x - 1$ spares or stand-by units to have a 99 per cent confidence that we shall not run out of spares. We can interpret this requirement that the reliability of the stand-by chain should be at least 0.99 for 20,000 hours of operation. Using Equation (12.2) with $m = 100{,}000$ and $t = 20{,}000$, we obtain

$$0.99 = (e^{-1/5})[1 + 1/5 + (1/5)^2(1/2) + \cdots + (1/5)^x(1/x!)]$$

Solving this equation for the nearest larger integer x which makes the right side equal to or larger than 0.99, we get

$$x = 3$$

Thus, having $x - 1 = 2$ spares to cope with chance failures, we would fulfill the requirement for a reliability of 0.99 and actually surpass it with a figure of 0.99885. By carrying a total of $10 + 2 = 12$ spares, we could be over 99 per cent confident that wearout and chance failures would not cause us to run short of replacements during a 20,000-hour space trip.

Chapter 23

SYSTEM RELIABILITY
MEASUREMENTS

THE METHODS OF MEASURING the reliability of components and assign
ing confidence limits to the obtained results explained in the preceding
two chapters apply equally well to measurements of system reliability.
However, in this chapter we shall discuss some specific details.

A system is a combination of components assembled to form a har-
monic whole and designed to perform one or more specific functions.
From the reliability-measurements point of view we distinguish two kinds
of systems: the *repairable* or *recoverable* system designed to perform for a
long time in a large number of missions or operations, and the *nonrepairable*, or more exactly, the *nonrecoverable* system designed to perform only
once.

The reliability of a system is basically determined by its design. However, the reliability of recoverable systems in the later stages of this operation depends on the maintenance policy, that is, on the overhaul, parts
replacement, and inspection schedules. In Chapter 7 we saw what a tremendous difference it makes whether preventive maintenance or only
repair maintenance is practiced. Repair maintenance waits until components in the system fail during the system's operation. It is the approach of replacing components only as and when they fail. Component
wearout then assumes an overwhelming role, as was shown in Figure 7.7,
and system failure rate stabilizes at a level much higher than it was at
the beginning. A newly built and debugged system is essentially free from
wearout failures and its reliability is governed by the frequency of chance
failures. When measuring the reliability of such a new system, the figure
which we get is not applicable after several thousand or even several

hundred hours of operation unless a maintenance policy is adopted which prevents wearout phenomena.

A closer inspection of Figure 7.7 reveals that it is only for a comparatively short period at the extreme left of the shown wearout failure frequency curve that we can expect wearout failures to be absent and only chance failures, if any, to occur. During this initial period preventive replacement of components is not necessary because it will not prevent chance failures. Therefore, components failing because of chance in this initial period are replaced only as they fail. In this period the system behaves exponentially with a constant system chance failure rate. After the initial period is over, component wearout begins to show up, and if only repair maintenance were practiced, system failure rate would rise drastically, then fluctuate, and finally settle to a comparatively very high constant value, as shown at the right end of the failure frequency curve. This final high but constant failure rate is predominantly caused by the wearout of components. In this final state the system again behaves exponentially, but with a much higher failure rate than its pure chance failure rate. Between the two extreme states, that is, the initial state when only chance failures occur and the final stabilized state, when wearout failures assume a constant rate which is superimposed on the chance failure rate, the system goes through a number of states with variable fluctuating failure rates. When high reliability is required, the purpose of preventive maintenance is never to allow a system to enter a condition in which wearout failures can appreciably affect system operation.

The determination of the times at which preventive maintenance actions are necessary is a matter of the study of the wearout characteristics of the various components which form a system. As a rule, various components will require various replacement times, which are found from component wearout tests described in Chapter 21. Such tests supply estimates of component mean wearout life and the standard deviation, and this information allows us to determine the optimum replacement or overhaul times for each type of components when the number of components of that type in the system is also considered.

When preventive maintenance is properly practiced so that it embraces all components known to be subject to wearout, a repairable system can operate in a pure chance failure rate condition for indefinitely long periods. Even after a large number of preventive and repair maintenance actions, such a system will still be as good as new. Thus, the concept of system longevity does not apply to preventively maintained systems. Strictly taken, there is no such thing as the mean wearout life of systems. When a system fails, it is usually not the end of its life, whereas when a nonrepairable component fails, its life is ended. With components we can say that if chance does not end the life of a component, then wearout will certainly end it. With systems we can only say that if no component chance failures occur in a system and the system therefore operates without failure for some longer period of time, there will come a moment when one of the components, if not preventively replaced, wears out and the system will fail—but this is not necessarily the end of the system's life.

Thus, if a system were required to operate for a long period of time without maintenance, and therefore were categorized as a "nonrecoverable" or "nonrepairable system," its chance failure rate would have to be extremely small. The probability that any of the system's components fails of wearout in the required maintenance-free period of operation would also have to be small. Because the wearout life of components is limited in most cases and also because the probabilities that individual components in the system will survive wearout must be multiplied by the product rule, the operating time limitations on so-called "maintenance-free" systems are very serious. In other words, long-life systems must be maintained preventively if long failure-free operating periods are required, and the system chance failure rate must be very low. As to the probability of the first failure of the system, we can ask the question: "Which is greater, the probability that a component chance failure will occur in the system or the probability that a component wearout failure will occur in the system?" The answer can be given only by means of a system probability analysis when the component parameters are known, and not by means of system tests.

For all practical purposes, system reliability measurements are limited to the exponential case—to the period when the system is new, possibly debugged, and before the components get a chance to fail of wearout. They are conducted on new systems, or at the various developmental stages of new systems, before service usage. Information about wearout is not obtained from these system measurements but must be obtained separately for the various components. In fact, component life information should be known before a system is designed because it determines what preventive maintenance schedules have to be prescribed. For nonrecoverable long-life systems which cannot be maintained during their long operation (space vehicles, etc.), the component wearout information determines the kind and amount of redundancy which has to be designed into a system to cope with wearout. A word of caution is necessary as to the kind of redundancy used. Obviously, so-called "parallel" redundancy gives practically no protection against wearout; stand-by redundancy must be used instead because two parallel components have about the same mean wearout life, whereas when two of them are in stand-by, the life of the combination is about doubled. Parallel redundancy is good only to avoid system chance failures.

As with components, reliability measurements of repairable systems consist of an estimation of the system's mean time between failures from a number of times between two successive system failures, obtained in tests. One or more equal systems are operated under simulated environmental conditions, the total accumulated operating time T is measured, and the number of chance failures r which occurred during the test is counted. The best and sufficient estimate of the system's mean time between failures is then obtained as

$$m = \frac{T}{r} \qquad (23.1)$$

If only a single system is under test, T is simply the straight operating time t_{op} of the system during the test. The idle periods when the system is shut off during the night or for repairs are excluded from the computation of the operating time. If the test on the single system is discontinued exactly at the time of the occurrence of the rth failure, the total accumulated operating time T is

$$T = t_{op} = \sum_{i=1}^{r} t_i \qquad (23.2)$$

where the times t_i are the operating times between two successive failures. Equation (22.18) is used to assign confidence limits on the system's mean time between failures. If, on the other hand, the test is discontinued at a preassigned time at which the system has operated for t_j hours since the last failure and is still operative at the time of test termination, the total accumulated operating time amounts to

$$T = t_{op} = \sum_{i=1}^{r} t_i + t_j \qquad (23.3)$$

In this case Equation (22.29) is used for confidence limits.

When n systems are operated simultaneously or in succession, the total operating time becomes

$$T = \sum^{n} t_{op} = \sum_{i=1}^{r} t_i + \sum_{j=1}^{n} t_j \qquad (23.4)$$

where t_{op} is the operating time of each system. The sum of the operating times of all systems then equals the term on the right of Equation (23.4) which consists of two sums, the first sum being the total of the accumulated operating times between failures t_i of all systems and the second sum being the total of the accumulated operating times t_j since the last failure of each individual system at the time of the test's termination. Obviously, when a simultaneous test on several systems is terminated, regardless of whether at the time of test termination one of the systems fails or not, the other systems will still be operative and each of them will have accumulated some operating time t_j since its last failure. These times should be added to obtain the total time T, and again Equation (22.29) should be used for confidence limits. However, when the number of observations is large—for instance, over twenty failures—Equation (22.18) and therefore also the graph in Figure 22.2 can be used regardless of the accumulated end times t_j. We would use Equation (22.29) in which the lower confidence limit is based on $2r + 2$ degrees of freedom only when the proportion of the Σt_j in the total observed operating time T exceeds, say, 10 per cent. Thus, when the number of failures is small. If no failures are observed during the time T, then

$$T = \Sigma t_j$$

and the proportion of the Σt_j in T is, in that case, 100 per cent. The times t_j can also be so-called "censored" times when one or more of the systems under observation or test are withdrawn from observation for any reason other than failure. For instance, when for the purpose of post factum reliability measurements a number of systems are being observed in actual service, their operating times being computed and failures counted, and we lose track of one or two of these systems during the observation period but we know that before we lost track of them, i.e., before they were withdrawn, some failure-free times since their last failures were observed on them, we would include those times in the total observed operating time T of all systems.

The above-described method of measuring, or rather, estimating the mean time between failures of a system applies to systems with exponential reliabilities. If the reliability function of a system is given by the exponential formula $R = \exp(-t/m)$, we have only to measure m which, being the single parameter in the formula, determines R for any operating time t. But systems often include redundancies which make the system reliability functions nonexponential. In this case the knowledge of the parameter m is not sufficient to determine system reliability.

Figure 23.1 shows the reliability function of a system containing a predominant amount of redundancy. The mean time between failures of this system can be estimated in the conventional way. It also equals the area under the reliability curve according to the equation

$$m = \int_0^\infty R(t)\, dt$$

and occurs at the time $t = m$ where the vertical line through m in the graph bisects the reliability curve in such a manner that area I equals area II. However, m does not determine the shape of the reliability curve because any number of nonexponential curves, including a particular exponential curve, can have the same mean time between failures and fulfill the condition that the areas I and II be equal. The dotted curve in the graph is an example. However, none of the other curves except the plain curve will satisfy the requirement that system reliability for an operating time of, say, t_1 hours be $R(t_1)$ as shown in the graph. This means that in order to prove that a system which contains redundancy has a required reliability of $R(t_1)$ for an operating time t_1, it is not enough to estimate its mean time between failures m; a test must be used which gives information about the actual shape of the system's reliability curve.

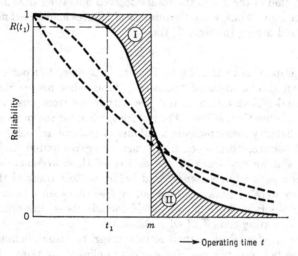

Fig. 23.1. Systems containing redundancy.

The technique used is based on the probability definition of reliability according to which the estimated reliability for any operating time t is given by

$$R(t) = \frac{S(t)}{N} \tag{23.5}$$

where N is the initial number of items subjected to a test and $S(t)$ is the number of items which did not fail up to time t, i.e., the number of surviving items. Assume, for illustrative purposes, that we place 100 equal systems under test at time $t = 0$ and count at regular intervals t_1, t_2, t_3, etc. the number of surviving systems $S(t_1)$, $S(t_2)$, $S(t_3)$, etc. We would have for the various operating periods the reliability estimates

$$R(t_1) = \frac{S(t_1)}{100}, \quad R(t_2) = \frac{S(t_2)}{100}, \quad R(t_3) = \frac{S(t_3)}{100}$$

etc. If the number of original systems were N, we would have N in the denominator instead of 100. We would continue this test until all original N systems fail, and from the reliability figures obtained for the various operating times we could graphically plot the reliability function of the system. This nonparametric technique yields a good approximation for the reliability function if N is large, and if the intervals t_1, t_2, etc. are chosen as small compared to the total test time.

Of course, in practice we would not use 100 systems, but we could do equally well with a single system and operate it until 100 or N failures are accumulated by repairing the system after each failure. With N failures we then have the same situation as if we had placed N systems originally under test and had let all of them fail one after the other without repairing them. However, now we must go through the trouble of measuring the times between each two successive failures, then order these times according to their length from the shortest to the longest one, and then use these times to count the number of "items" failing up to an operating time t, or surviving t. To speed up the test we can operate several systems and repair each after it fails, measuring on each system the operating time between two successive failures of that system. This will more quickly generate the needed N operating times between failures. When we have ordered these N operating times between failures according to their length, we evaluate the system reliability $R(t)$ for system operating time t by counting as "surviving items" the number of the ordered operating times between failures which are longer than t. The reliability estimate for a system operating time t is again given by Equation (23.5); however, $S(t)$ becomes the number of operating times between failures which were longer than t, and N is the total number of these times between failures obtained in the test and is simply equal to the total number of observed failures. Two methods based on this technique have been described in detail, one by G. R. Herd* and the other by S. R. Calabro and S. Pearlman.†

When using this technique of estimating the reliability function of systems which contain redundancy, it is essential that failed redundant components not be replaced until the entire redundant arrangement fails so that it causes the system to fail. When this happens, then all the failed components in the redundant arrangement are replaced and the test can be resumed. In other words, no preventive maintenance must be used during the test if we want to explore the entire reliability function

* G. R. Herd, "Estimation of Reliability Functions," ARINC Monograph No. 3 (May 1, 1956).

† S. R. Calabro and S. Pearlman, "Simulation Techniques Verify Reliability," *Proceedings* of the Third National Symposium on Reliability and Quality Control (January 1957), p. 216.

of a system. Only in this way can we get information about the system's capability or probability to operate in one stretch without failure for any operating time t. We simultaneously also obtain information about the system's mean time between failures when it is operated without preventive maintenance.

On the other hand, it is often required that a system have a specified reliability $R(t_1)$ for an exactly specified operating time t_1. To measure this reliability we could take N systems again and operate them simultaneously for the predetermined time t_1 and count the number of systems which survive this test. If 99 per cent of the systems survive the time t_1, the system reliability estimate for t_1 is 0.99. Confidence limits can now be assigned on the binomial distribution of R, as in quality control tests. If the system contains redundancy, the measured estimate $R(t_1)$ is valid only for the operating condition that at the beginning of the t_1 mission time all components in the redundant arrangements be good. If the system is repairable and scheduled for a large number of successive operations or missions of t_1 hours each, the measured estimate $R(t_1)$ will be good only for those missions which begin with all redundant components in good condition and with all system components operating in their useful life period; that is, no component must have approached a wearout state.

Naturally, we again can make the test for $R(t_1)$ with a single system and operate it N times for t_1 operating periods. System estimated reliability $R(t_1)$ is then given by the percentage of times the system has operated for t_1 hours without failure. To generate the information faster, we can take two or more systems. When we make N attempts to operate the system without failure for t_1 hours and we succeed in S attempts, the system reliability estimate is $R(t_1) = S/N$. We see that the number of systems with which we make the N attempts is immaterial, although it is recommended to test more than one system if possible to be assured of the uniformity of the production process.

If a definite maintenance policy is prescribed, such as checking redundant systems for failed components after every ten t_1 operations of a system, and/or preventively replacing components which are subject to wearout after every x hours of system operation, such requirements can be incorporated in reliability tests but the tests become very long.

The important thing to remember about reliability measurements is that we must define exactly for which system condition the measurement is to be made and realize that, according to the requirements and system design, the measurement procedures have to be tailored individually for each specific case.

A special case of reliability measurements are sequential reliability tests used when we are not interested in the actual reliability of a system or equipment and only want to be assured that the reliability at a given confidence level is better than a specified minimum value. We shall deal with these tests in the following chapter.

However, before concluding this chapter we want to mention so-called "system maintainability tests." These are performed by repair technicians skilled similarly to service maintenance crews and consist of removing, repairing, reinstalling, or replacing certain specified parts in the system and measuring in manhours the time required for such operations. This information can be used for the calculations described in Chapter 17.

Chapter 24

SEQUENTIAL
RELIABILITY TESTS

ALTHOUGH THE PURPOSE of the various methods of reliability measurements in Chapters 21 and 23 is to obtain estimates of the true value of reliability of components and systems at chosen confidence levels, the purpose of reliability tests is to establish in the shortest possible test time and at minimum cost whether or not the reliability of a type of components or of a system is equal to or better than a specified minimum. Thus, in reliability tests we are not interested in the actual reliability of the tested equipment, but rather in the proof that the equipment is at least as good as or better than required. To accomplish this the method of sequential probability ratio tests can be used.*

Characteristic of this method is that the number of observations— in our case, the number of observed times between failures, or, which is the same, the number of actually observed failures—is not predetermined but depends at any instant on the outcome of the preceding observation. The method supplies an exact rule for making one of three decisions at any instant during the test: (1) accept, (2) reject, (3) continue testing. Which of these three decisions is made depends on the outcome of the observation immediately preceding the time at which a decision is being made. If the rule shows that decision (1) or (2) has to be made, the test is terminated. If it shows that decision (3) must be made, the test continues to obtain more information, such as more failure-free operating

---

* This method was devised in 1943 by A. Wald and is described in his paper "Sequential Tests of Statistical Hypotheses," *Annals of Mathematical Statistics*, XVI.2, June 1945, pp. 117–186.

time or an additional failure; this is why the test is called "sequential." In the following we shall apply this method to events which occur in the time domain, such as chance and wearout failures.

The sequential probability ratio test can be explained as follows: If m is the true parameter point of a statistical distribution, where m can be the single parameter of a distribution or a parameter point determined by the coordinates of a number of parameters a, b, c, etc., the hypothesis H_0 that $m = m_0$ is tested against the hypothesis H_1 that $m = m_1$. The value of m_1 is the specified minimum acceptable value of the parameter, and m_0 is some chosen upper value such that $m_0 > m_1$. The probability that r observations will be made in test time t or that r failures will occur in test time t is $P(r/m)$ for the true parameter m. For the specified minimum acceptable value m_1 of the parameter, the probability that r observations will be made in time t is $P_1(r/m_1)$, and for the arbitrarily chosen upper value m_0 this probability is $P_0(r/m_0)$. Because m_1 is specified and m_0 is chosen in advance of the test, the values P_1 and P_0 can be calculated for an assumed distribution (for instance, exponential or normal), and compared with the observed results available at any stage of the test, i.e., at any test time t by which time r failures have been observed on the tested equipment. This is done so that at each stage of the test the probability ratio P_1/P_0 is computed and compared against two selected positive constants A and B which are determined from the prescribed strength of the test, i.e., from the risks agreed upon between the consumer and the producer—the consumer's risk β of accepting equipment with a parameter smaller than the specified minimum m_1 and the producer's risk α of rejecting equipment with a parameter larger than the chosen upper value m_0.

The constants A and B are approximately given by*

$$A = \frac{1 - \beta}{\alpha} \tag{24.1}$$

$$B = \frac{\beta}{1 - \alpha} \tag{24.2}$$

Because β is the test's probability of accepting bad equipment, or rather, the risk which the consumer is prepared to take, $1 - \beta$ is the probability of rejecting bad equipment. Similarly, α being the test's probability of rejecting good equipment, or the risk which the producer is prepared to take, $1 - \alpha$ is the probability of accepting good equipment. The constant A is the ratio of the probability of rejecting bad equipment to the probability of rejecting good equipment; A is usually considerably larger

---

* A. Wald, *Sequential Analysis*, John Wiley and Sons, Inc., New York, (1947), pp. 40–48.

than unity. The constant B, on the other hand, is the ratio of the probability of accepting bad equipment to the probability of accepting good equipment, and is smaller than unity. As a rule, therefore, $B < A$.

As the test proceeds step by step, the probability ratio P_1/P_0 is continuously being computed from the test results available at that instant or stage and compared with the agreed constant probability ratios A and B. If at any stage of the test the computed probability ratio P_1/P_0 is equal to or smaller than B, an "accept" decision is made with regard to the tested equipment. If P_1/P_0 is equal to or larger than A, a "reject" decision is made, and if P_1/P_0 lies between A and B the test is continued to gain more information.

The test rule is therefore as follows:*

1. Accept if: $$\frac{P_1}{P_0} \leqq B \qquad (24.3)$$

2. Reject if: $$\frac{P_1}{P_0} \geqq A \qquad (24.4)$$

3. Continue testing if: $$B < \frac{P_1}{P_0} < A \qquad (24.5)$$

If the test is continued long enough, it will terminate with an accept or reject decision. If the equipment is much better or much worse than the specified minimum, a decision will generally be reached very soon. The important conclusion which can be drawn from the final result is that if an accept decision has been made according to rule 1, the true parameter m exceeds the specified minimum m_1 with a probability of at least $1 - \beta$. This conclusion can also be interpreted as a $100(1 - \beta)$ per cent minimum confidence level that m exceeds m_1. The actual confidence level is, however, much higher. On the other hand, if a reject decision has been made according to rule 2, there is a probability of at least $1 - \alpha$ that the true parameter m is smaller than the arbitrarily chosen upper value m_0.

The advance choice of the probabilities α and β is a necessary condition for conducting sequential probability ratio tests. These probabilities determine in advance the strength of the test, i.e., the maximum acceptable risks of obtaining wrong decisions when accept or reject decisions are made. As we know, all statistical tests contain some probability of giving wrong decisions; there is nothing unusual in asking that the probabilities α and β of getting wrong decisions be limited in advance by choice or agreement.

Apart from determining the risks, the choice of α and β has an effect on the test time. The smaller α and β are chosen, the longer the test will

---

* See Wald, *Sequential Analysis*, pp. 37 and 38.

last with equipment of equal reliability. As a compromise, it has been recommended to choose $\alpha = \beta = 0.1$, or 10 per cent, for electronic equipment.* We have then $A = 0.9/0.1 = 9$, and $B = 0.1/0.9 = 0.111$. As to the choice of the upper limit m_0, this also has a bearing on the test time. The closer m_0 is chosen to the specified minimum m_1, the longer the test will last. Thus, in the interest of reducing test time and cost, m_0 should be chosen considerably larger than m_1. The risk of rejecting equipment better than the chosen upper m_0 is not affected by the choice of m_0 and depends only on the choice of α. But the probability that equipment worse than m_0, though better than m_1, will be rejected increases when m_0 is chosen high.

Obviously, if m_0 were chosen very large, the concept of the producer's risk α would lose its meaning if the producer knew in advance that the true m of the equipment which he built can never reach m_0. As a matter of fact, he has advance information available from the design reliability analysis. The choice of m_0 should therefore be left to the producer.

When α or β is chosen high, Equations (24.1) and (24.2) from which A and B are computed are no longer valid because they are only approximations for the constants A and B when α and β are small. The exact determination of the constants A and B is a very complicated procedure. To make use of the approximations, α and β should be kept small.

The actual numerical choice of α, β, and m_0 depends on the available test time and on the number of equipments which can be tested. More equipments will generate more quickly the information required to make an accept or reject decision because it is the total operating time accumulated together by all the equipments under test which decides how soon a decision will be reached.

From the consumer's point of view, the two important figures in the test are the specified minimum acceptable m_1 and the risk β of accepting equipment with a true parameter m smaller than m_1. Of course, the consumer is interested in minimizing this risk. He would like to reduce it to 0.05, or even 0.01. However, this would necessarily increase test duration and cost. As to the specified minimum m_1, the question arises of what is the minimum figure to be used in a particular case. Obviously, m_1 must not be specified smaller than required in the actual operation of the equipment. If experience indicates that a deterioration of equipment reliability takes place between the time the equipment is produced and the time it goes into operation, an empirical deterioration factor $K < 1$ must be considered in specifying m_1. Thus, if we know that an equipment which has a parameter m_a when leaving the production line and which in operation

---

* See *Reliability of Military Electronic Equipment*, Report of Advisory Group on Reliability of Electronic Equipment, Office of the Assistant Secretary of Defense (Research and Development), Washington, 4 June 1957, pp. 167 and 191.

will display a parameter Km_a, and Km_a is the minimum acceptable figure for operation, we shall specify m_a as the minimum contractual parameter to be proven in test, and not Km_a. In other words, if the minimum needed reliability of an equipment in operation is given by a parameter m but experience has proven that a $100(1 - K)$ per cent deterioration occurs in the parameter after testing the equipment in the producer's laboratory, we would require the producer to build equipment with a parameter

$$m_1 = \frac{m}{K} \qquad (24.6)$$

and would specify that the producer has to prove by test that the true parameter of the equipment is equal to or larger than m_1 at a specified consumer's risk level β.

The effects which the choice of α, β, and m_0 have on a sequential probability test will become clear as we derive the formulas for the exponential case.

From Poisson's equation we know that an exponential equipment which has a true but unknown mean time between failures m will have a probability of failing r times in an accumulated operating time t of

$$P(r) = \left(\frac{t}{m}\right)^r \frac{e^{-t/m}}{r!} \qquad (24.7)$$

We want to find in a sequential test whether the true reliability $R = e^{-t/m}$ of an equipment is at least equal to or better than a specified minimum acceptable reliability $R_1 = e^{-t/m_1}$. Thus, the test should prove that m is at least equal to or larger than m_1. Obviously, if the true mean time between failures of the equipment were exactly the specified minimum m_1, the equipment's probability of failing r times in the accumulated operating time t would be

$$P_1(r) = \left(\frac{t}{m_1}\right)^r \frac{e^{-t/m_1}}{r!} \qquad (24.8)$$

To be able to conduct this test we choose an arbitrary upper value m_0. If the equipment's mean time between failures were exactly this upper value, we get for the probability of r failures

$$P_0(r) = \left(\frac{t}{m_0}\right)^r \frac{e^{-t/m_0}}{r!} \qquad (24.9)$$

We now form the probability ratio of (24.8) and (24.9),

$$p(r) = \frac{P_1(r)}{P_0(r)} = \left(\frac{m_0}{m_1}\right)^r e^{-[(1/m_1)-(1/m_0)]t} \qquad (24.10)$$

and compare this ratio at any stage of the test against two chosen constants A and B given by Equations (24.1) and (24.2). Obviously, at any accumulated operating test time t, when r observations (failures) are available, we can instantaneously compute the probability ratio $p(r)$ because we know m_1 and m_0 as well as t and r at any instant during the test. Now if at any instant $p(r)$ becomes equal to or smaller than B, we immediately make an accept decision; if $p(r)$ becomes equal to or larger than A we immediately make a reject decision. If the computation at time t shows that

$$B < p(r) < A \qquad (24.11)$$

we continue testing until $p(r)$ hits either A or B.

Once m_1 has been specified and α, β, and m_0 chosen or agreed upon, Equations (24.10) and (24.11) are all we need for a numerical conduction of the test. Let us assume that $\alpha = \beta = 0.1$, so that $B = 0.111$ and $A = 9$, and that $m_1 = 100$ hours and we choose $m_0 = 2m_1 = 200$ hours. Equation (24.10) then assumes the form

$$p(r) = (2)^r e^{-t/200} \qquad (24.12)$$

Thus, in this particular case, when up to an accumulated operating time t the equipment has failed r times, we obtain $p(r)$ by forming the rth power of 2 and multiplying this by $e^{-t/200}$ obtained from exponential tables. This computation of $p(r)$ must be made in comparatively short intervals during the test so as to be sure that the test is terminated as soon as $p(r)$ becomes equal to A or B. The test requires continuous attention. If no failure has occurred up to an accumulated 200 hours, $p(r)$ at 200 hours equals $(2)^0 \times e^{-1} = 0.368$. The test must obviously continue because this value lies between $B = 0.111$ and $A = 9$, but it is quite close to B. To become B, the test would have to continue without failure until $e^{-t/200}$ becomes 0.111, which occurs at an accumulated operating time of $t = 440$ hours, at which time the test is terminated with an accept decision. The minimum accumulated time required by the test for an accept decision when $\alpha = \beta = 0.1$ and when $m_0 = 2m_1$ is

$$T(0)_{min} = 4.4m_1 \qquad (24.13)$$

if no failure occurs and if only a single equipment is tested. If n equipments are tested and no failure occurs, the time T_{min} is still the minimum required accumulated operating time to reach an accept decision. But T_{min} is the time accumulated by all n equipments and therefore in terms of straight test time t we obtain

$$t(0)_{min} = \frac{T(0)_{min}}{n} = \frac{4.4m_1}{n} \qquad (24.14)$$

With five equipments, if no failure occurs, the accept decision is made in 88 hours of straight test time by which time we have accumulated 440 unit hours of operation without failure on the five sets together, which is sufficient for an accept decision according to (24.12) because at that moment $p(r) = 0.111$. We are then justified in making the statement that there is a probability of at least 90 per cent that the true m of the equipment is larger than or at least equal to 100 hours. In fact, this probability is 98.7 per cent. The test thus yields results at a very high confidence level.

In the case of a single equipment, if the equipment were to fail before it had operated for 440 hours, the probability ratio at 440 hours would be $(2)^1 \times e^{-44\%00} = 2 \times 0.111 = 0.222$. The test would have to continue without a second failure until $e^{-t/200}$ becomes one-half of 0.111, i.e., 0.0555, because $2 \times 0.0555 = 0.111$ and this occurs at $t = 580$ hours, in which, of course, the time required to repair the first and only failure must not be included. For an accept decision a single equipment would have to accumulate $T(1)_{\min} = 580$ hours of operating test time if one failure occurred before 440 hours and no failure afterwards.

In the case of n equipments, if one of them failed before all of them together have accumulated 440 hours, the test would have to continue until 580 hours are accumulated without a second failure. The figure of 580 hours is not affected by whether or not the failed equipment is repaired, but if the test continues with only $n - 1$ equipments it will take somewhat longer, as measured in straight clock time, to accumulate the required 580 hours.

From this discussion it is clear that in sequential tests, as in many other reliability tests discussed in the preceding chapters, we must measure the accumulated operating time, i.e., the sum of the operating times of all the equipments under test.

If two, three, or more failures occur during the test before an accept decision becomes possible, the required minimum accumulated operating time $T(2)_{\min}$, $T(3)_{\min}$, etc. required to reach an accept decision becomes longer and longer.

Now let us look at what happens with our example if the equipment is worse than required. When can a reject decision be made? Obviously, the value of $p(r)$ as given by Equation (24.12) must become equal to or larger than $A = 9$ to make a reject decision. If the equipment fails three times in succession right at the beginning of the test, so that for all practical purposes $t = 0$ and $e^{-t/200} = 1$, $p(r)$ will be $(2)^3 = 8$. According to the test rules, this does not yet justify a reject decision because A has been made a high figure (i.e., 9) by the choice of the consumer's and producer's risks β and α. But if four failures are observed before an operating time of 116 hours is accumulated, a reject decision is made immediately because $(2)^4 = 16$ and $e^{-t/200} = \%_6 = 0.56$ which corresponds to 116 hours, i.e., $e^{-11\%00} = 0.56$. Again, it does not matter whether one or n equipments are under test, except that the 116 unit hours are accumulated faster with more equipments tested.

The general equations for the minimum required accumulated operating time in a sequential test are

$$\text{For accept:} \quad T_{min} = \frac{\ln B + r \ln (m_1/m_0)}{(m_1 - m_0)/(m_1 m_0)} \tag{24.15}$$

$$\text{For reject:} \quad T_{min} = \frac{\ln A + r \ln (m_1/m_0)}{(m_1 - m_0)/(m_1 m_0)} \tag{24.16}$$

These times are independent of the number of equipments tested because they are the operating times in unit hours accumulated by all the tested equipments together. Their meaning is that a sequential test in which A, B, m_1, and m_0 have been fixed cannot terminate faster with an accept or reject decision than in T_{min} accumulated operating time. Of course, it can last much longer when, as the test progresses, $p(r)$ continues to move between A and B.

The above equations clearly show that the test time of an equipment with a certain true m depends on the choice of A and B, and therefore on the choice of α and β as well as on the choice of m_0. Returning to our example (24.12), if we were to make $m_0 = 10m_1$ we would obtain

$$p(r) = (10)^r e^{-9t/1000} \tag{24.17}$$

A reject decision would be made now if even a single failure occurred at any time prior to 11 unit hours of accumulated operating time, because $9 = (10)^1 e^{-9t/1000}$ which gives $t = 11$ hours. However, the only statement we can now make about a reject decision is that with a probability of $1 - \alpha = 0.9$ the equipment's m is worse than $10m_1 = 1000$ hours. Thus, it could happen that we unnecessarily reject good equipments as a possible sacrifice for reducing the test time. An accept decision with no failures having occurred would be made if an operating time of 245 hours is accumulated, because $0.111 = (10)^0 e^{-9t/1000}$ and $t = 245$. If one failure occurred prior to the accumulation of 245 hours, the accept decision would have to wait until an operating time of 500 hours was accumulated without a second failure's occurring, because $0.111 = (10)^1 e^{-9t/1000}$ and $t = 500$. When an accept decision has been reached at any stage of the test, we would state that the true m of the equipment is at least 100 hours with a probability of at least 0.9.* Thus, because m_1 was not changed, the probability statement concerning an accept decision remains as before for the same 100-hour level.

* Confidence levels for sequential tests are discussed at the end of this chapter.

Comparing the test times with the previous case when m_0 was chosen to be $2m_1$, we see that by increasing m_0 to $10m_1$ we speed up testing; in particular, we speed up reject decisions because these refer to the m_0 level. The speed of accept decisions is only slightly increased. To increase this speed β would have to be made larger than 0.1, but this increases the risk of accepting equipment with an m smaller than m_1. An increase of the acceptance speed can be obtained by increasing α from 0.1 to, say, 0.4; however, this seriously increases the risk of rejecting good equipment. If equipment is rejected in the test, the statement could be made that there is a probability of at least $1 - \alpha = 0.6$ that the true m of the equipment is smaller than m_0. Obviously such increases of m_0 to $10m_1$ and of 0.1 to 0.4 could be acceptable to a producer only if he were very sure that the equipment which he is building and which must have a minimum m_1 of 100 hours has a true m greatly in excess of 1000 hours. Such cases, however, do happen and test time and cost can be reduced substantially.

Thus, although for normal procedures $\alpha = \beta = 0.1$ and $m_0 = 2m_1$ seem to be a good choice, when time is at stake or when a good indication of the true reliability of the equipment to be tested exists, these figures can be considerably modified from case to case according to the prevailing conditions, giving careful consideration to the effects of such modifications. No general rules can be given.

If with $\alpha = \beta = 0.1$ and $m_0 = 2m_1$ the minimum test times $T_{\min}$ in Equations (24.15) and (24.16) are acceptable from the point of view of available time and available number of equipments for the test, but it is necessary to reduce the additional test time if no accept or reject decisions result some time after $T_{\min}$ has elapsed, the procedure of test truncation can be used. However, we must realize that test truncation will affect both α and β, according to the number of observations available at the time of truncation. Thus, truncation affects test precision. The following rule for test truncation is given:[*]

$$\text{Accept if:} \quad \ln B < \ln p(r) \leq 0 \qquad (24.18)$$

$$\text{Reject if:} \quad 0 < \ln p(r) < \ln A \qquad (24.19)$$

The value of $p(r)$ in the above equations is that computed for the instant when decision to truncate is being made. Thus, a test can be specified to proceed normally until t operating hours are accumulated together by all the equipments under test, where t exceeds the larger one of the two $T_{\min}$ figures given by Equations (24.15) and (24.16); if no reject or accept decision has been made by time t, but the rule shows that the test should continue, the test is truncated and a decision made for accept or reject

---

[*] See A. Wald, *Sequential Analysis*, p. 61.

according to Equation (24.18) or (24.19), whichever applies. If truncation is made early after T_{min} has been accumulated, the probabilities α and β may be affected very significantly.*

In cases when the specified minimum mean time between failures m_1 of the equipment is very long (thousands of hours or even tens of thousands of hours, as for space equipment), reliability tests become a real problem. Obviously, a large number of equipments would be needed for such tests to accumulate the required operating time within a reasonable calendar time for an accept or reject decision. However, usually only very few equipments will be available for test because only a few of them have to be built altogether. The best procedure is to have available reliable information on the reliability measurements of the components from which the equipment is built, to have the equipment undergo a series of functional performance tests under simulated operating conditions, and to rely on a detailed reliability analysis of the equipment, taking great care to properly evaluate the stress levels at which the individual components will operate when the equipment is placed into service. If one or two of the equipments still have to undergo reliability tests, it becomes necessary to considerably relax the requirements for α and β and choose a larger m_0; test truncation may still be necessary to complete the test within a foreseeable number of months. Also, care must be taken not to operate any equipment in reliability tests beyond the period at which wearout of components may occur.

To facilitate the conduction of sequential reliability tests, we shall now show the conventional graphical method. This method is very useful because it eliminates the necessity of continuously computing the probability ratio $p(r) = P_1(r)/P_0(r)$ once the test has been properly prepared.

We start by substituting for $p(r)$ in Equation (24.11) the expression given in (24.10). Taking the natural logarithm ln for all terms we obtain

$$\ln B < r \ln \frac{m_0}{m_1} + \left(\frac{1}{m_0} - \frac{1}{m_1}\right) t < \ln A \qquad (24.20)$$

We transform this inequality by dividing all terms by $\ln (m_0/m_1)$ and subtracting from each term $(1/m_0 - 1/m_1)t$ or, which is the same, by adding to each term $(1/m_1 - 1/m_0)t$:

$$\frac{\ln B}{\ln (m_0/m_1)} + \frac{(1/m_1 - 1/m_0)}{\ln (m_0/m_1)} t < r < \frac{\ln A}{\ln (m_0/m_1)} + \frac{(1/m_1 - 1/m_0)}{\ln (m_0/m_1)} t \qquad (24.21)$$

We see from this equation that if r, the number of actually observed failures at any stage of the test, lies numerically between the two values on

* For an evaluation of the effects of truncation on α and β, see Wald, *Sequential Analysis*, chap. 3.8.

the left and on the right side of the unequality, the test must continue. If at any stage r becomes equal to or smaller than the left side, the test is terminated with an accept decision; if r becomes equal to or larger than the right side, the test is terminated with a reject decision.

The inequality given by Equation (24.21) can also be written in the following form:

$$a + bt < r < c + bt \tag{24.22}$$

where the left and right sides are equations of two parallel straight lines with equal slopes b. When these two lines are plotted on paper with t as abscissa and r as ordinate, the constants a and c are the intercepts of the lines with the ordinate. The numerical computation of a, c, and b is given by

$$a = \frac{\ln B}{\ln (m_0/m_1)}$$

$$c = \frac{\ln A}{\ln (m_0/m_1)} \tag{24.23}$$

$$b = \frac{(1/m_1 - 1/m_0)}{\ln (m_0/m_1)}$$

The constants c and b will always be positive, whereas a will be negative because B is always smaller than unity. The two parallel lines $a + bt$ and $c + bt$ are then accurately plotted on graph paper in an rt coordinate system as shown in Figure 24.1; between these two lines we plot the step function representing the cumulative number of failures r vs. the time t. The two parallel lines divide the rt space into three regions: the accept region at the bottom below the $a + bt$ line, the reject region on the top above the $c + bt$ line, and the region of undecision, or the region of "continue testing," between the two lines.*

It is important to realize that the time t on the abscissa in Figure 24.1 is not the straight clock time, but is the operating time in unit hours accumulated together by all the equipments under test up to any instant. Only when a single equipment is under test will t equal the active test time t_a as measured on the clock. Of course, t can be converted into t_a if n equipments are tested, so that $t_a = t/n$, but this is valid only if so-called

---

* A graphical method for sequential reliability tests in which failed test-specimens are immediately replaced is described by B. Epstein and M. Sobel in their pioneering paper "Sequential Life Tests in the Exponential Case," *Annals of Mathematical Statistics*, 26.1, March 1955, pp. 82–93. A method which also considers non-replacement is shown in the paper "Acceptance Testing under the Assumption of Constant Failure Rate" by Judah Rosenblatt, Consultation Note No. 11, Boeing Airplane Company, Renton, Wash., July 17, 1957. In our treatment we plot the accumulated unit-hours on the abscissa, resulting in a graph which is applicable to both replacement and non-replacement tests.

"immediate replacement" of failed equipments takes place so that exactly n equipments are being tested at all times. In the majority of cases, however, a failed equipment must be repaired and during the repair the test continues with $n - 1$ equipments. Two equipments may also be out of

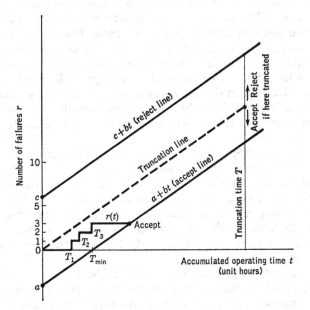

Fig. 24.1. Sequential reliability test for $\beta > \alpha$.

the test for repair at the same time. The time conversion formula therefore reads

$$t_a = \frac{t}{n} - \Sigma \, t_{out} \tag{24.24}$$

where the sum of the t_{out} times is the total idle time accumulated up to t_a by the equipments which for any reason have missed certain portions of the test. Because time measurements must be made in any case, it is more practical to use the total accumulated unit hours t on the test's time scale.

The parallel dashed line in the graph, which goes through the point of origin, represents the truncation line. If it is specified that the test be truncated at some time T if no prior accept or reject decision has been possible, the truncation rule says that an accept decision is made if at T the $r(T)$ value of the $r(t)$ step function is below the dashed line or touches it, and a reject decision is made if at T the value $r(T)$ is above the dashed line. If prior to the truncation time a comparatively large number of observations has been made, i.e., if T is large with respect to the minimum time T_{min} for making an accept decision, the effect of the truncation on α and β will be small.

It can be seen from the graph that T_{min} for acceptance occurs at the intersect of the lower $a + bt$ line with the t axis. This time T_{min} can be exactly calculated from Equation (24.15) for $r = 0$, and this point can be used for an accurate plotting of the two lines. Thus, when no failures occur up to T_{min}, this is the earliest time an accept decision can be made. A calculation of T_{min} will also help to scale properly the abscissa on the graph. As to scaling the ordinate, the intercept of the $c + bt$ line with the r axis, which equals the constant c as given in (24.23), determines the number of failures which would have to occur immediately after the test is started to make a reject decision. Thus, c helps to place the proper r scale on the ordinate.

With the test prepared as described, all that remains is to compute at intervals—the best choice is when failures occur, and when equipments are withdrawn from or returned to the test—the total accumulated operating time in unit hours and plot the step function r vs. t with the steps entered in the graph at the times T_1, T_2, T_3, etc. when the failures occurred. According to the rules, the test terminates as soon as an accept or reject decision can be made, or at the latest, at the prescribed truncation time T.

Looking at the equations for a, c, and b in (24.23), we can see immediately that a sequential test in which we specify that m_0 should be m_1 becomes impossible, because $\ln 1 = 0$. When m_0 is very close to m_1, the points a and c will be far apart, the area of indecision between the lines $a + bt$ and $c + bt$ becomes wide, and it takes a long time to arrive at a decision. The farther m_0 is from m_1, the closer become the points a and c and the two lines, and the faster a decision can be reached because $r(t)$ will hit one of the two lines sooner. Also, as m_0 gets larger relative to m_1, the slope b of the two lines decreases. Thus, with increasing m_0 the two lines are coming closer together, both are approaching the truncation line, and the slope of the whole graph decreases. In the limit, as m_0 increased to infinity, the two lines would coincide with the abscissa and no sequential test is possible. The two lines would also coincide when $A = B$, i.e., when $\alpha = \beta = 0.5$. We would then have only one line with a slope b, and a sequential test is again impossible. Thus, we cannot specify that both the consumer's risk and the producer's risk should each be 50 per cent. On the other hand, the smaller α and β become, the farther will the points a and c, and therefore the two lines, shift from each other. When only β is made larger, B increases and A decreases; therefore both points a and c shift closer to the point of origin O. However, a approaches O relatively faster than c. Thus, we get easier accept decisions than reject decisions for the same equipment. When only α is made larger, c approaches O relatively faster than a, and we get easier reject decisions for the same equipment.

Equations (24.23) assume a particularly simple form in the special case when we choose $m_0 = 2.7183m_1$, because m_0/m_1 then equals the base of the natural logarithm, which is $e = 2.7183$, and $\ln(m_0/m_1) = \ln e = 1$. Therefore, all the denominators become unity, and we obtain in this special case

$$a = \ln B, \qquad c = \ln A$$

and

$$b = \frac{1}{m_1} - \frac{1}{m_0} = \frac{1}{m_1} - \frac{1}{em_1} = \frac{e-1}{em_1} = \frac{0.63}{m_1} = \frac{1}{1.59m_1}$$

We can make the slope b to take any angle against the abscissa in the graph by proper choice of scales. For instance, if we wish the two lines to be inclined at an angle of 45 deg in the above special case, we have to choose one division on the ordinate to represent one failure and one division of the same length on the abscissa to represent an accumulated operating time t of $1.59m_1$ unit hours. Then the slope $b = 1/1.59m_1$ will have an angle of 45 deg on the graph. But we get better time readings if we expand the time scale on the ordinate, which naturally decreases the graph angle although b remains the same. Thus, if we were to take one division equal to $0.5m_1$ the angle would be about 20 degrees.

The points a and c on the ordinate are calculated directly in units of r. With $\alpha = \beta = 0.1$, we have $A = 9$ and $B = \frac{1}{9} = 0.111$. Therefore, in the example above

$$c = \ln 9 = +2.19722 \quad \text{and} \quad a = \ln \frac{1}{9} = -2.19722$$

Or, if we take $\alpha = \beta = 0.2$, we get

$$A = 4, \quad B = \frac{1}{4}, \quad c = +1.38629, \quad \text{and} \quad a = -1.38629$$

As we see, whenever $\alpha = \beta$, the points a and c on the ordinate lie at equal distance but in opposite directions from the point of origin O.

The graph for a sequential reliability test should always be prepared in advance. It helps to clarify between the producer and consumer points which affect the test parameters. The preparation of the graph itself is easy and it considerably simplifies the test procedure because no further elaborate calculations are necessary. The problem is reduced to computating the accumulated unit hours and entering in the graph the cumulative $r(t)$ line as the test proceeds.

The test described above is an exponential test because the equations for the accept and reject lines are derived from the exponential function. The test assumes a constant failure rate and can be applied only when this assumption is correct. Thus, the test is good for debugged equipment in which no wearout failures will occur during testing. If an accept decision is made, it tells us that the exponential true mean time between failures of the equipment is larger than m_1; if a reject decision is made, it tells us that the true m is smaller than m_0.

Because the graph refers to only those failures which occur at constant rates, i.e., to chance failures, we would not include in the graph failures which are not of a chance nature if such should occur during the test, nor would we include so-called "secondary" or "dependent" failures which occur simultaneously as a consequence of a primary chance failure.

A reliability test requires a continuous monitoring of the equipment or system performance characteristics and their comparison with the performance limits prescribed in the specification. This continuous monitoring is necessary because it helps to establish out-of-tolerance malfunctions which, from the reliability point of view, are failures in the same category as faults causing complete stoppage. Such malfunctions as well as complete failures must be entered in the graph except when they can be attributed to

1. testing errors, errors in instrument readings, or faults or equipment damage caused by the test personnel;
2. manufacturing errors, such as improper wiring, use of incorrect parts, material faults, etc. which can be corrected, for instance, by debugging procedures or by stricter quality control and production inspection so that they will not recur in service or in other lots to be shipped;
3. secondary failures, caused directly by primary chance failures of other components or by failures of auxiliary equipment, such as external power supply failures;
4. maladjustments, which can be corrected during normal operation without the use of test equipment or tools, i.e., when provisions for adjustment by the operating personnel are built into the equipment so that normal operation does not need to be interrupted. However, if maladjustments require interruption of operation to be corrected, they must be considered as failures and included in the graph.

During the test a failure log* should be maintained in which all failures and malfunctions, including the exempt ones, are entered and identified by the parts involved, nature of the cause, category of the failure, and time of occurrence. The failure log helps to check whether the test

---

* For recommended logging procedures in reliability tests see *Reliability of Military Electronic Equipment*, pp. 115–208.

was correctly performed, whether the appropriate types of failures were entered in the graph, and gives valuable information as to the corrective actions which have to be taken in production to eliminate the occurrence of other than chance failures—or even in design if the chance failure rate is too high. As we know, the frequency of chance failures—and therefore equipment reliability—is basically decided during the design stage, but careless production and assembly can introduce a multitude of other failures which reduce reliability below the design level.

Sequential probability ratio tests based on Equations (24.3), (24.4), and (24.5) also can be designed for distributions other than exponential. Of course, the probability ratio $p(r) = P_1(r)/P_0(r)$ is then no longer given by Equation (24.10) and has to be calculated from the particular distribution involved. When this has been done, graphical procedures can again be applied as we did in Equations (24.20) and (24.21) for the exponential case. For instance, to test whether the mean wearout life M of a component population exceeds a specified numerical value, Equation (24.21) assumes the form

$$\frac{2 \log B}{M_0^2 - M_1^2} - \frac{2(M_1 - M_0) \overset{r}{\Sigma} T_i}{M_0^2 - M_1^2} < r < \frac{2 \log A}{M_0^2 - M_1^2} - \frac{2(M_1 - M_0) \overset{r}{\Sigma} T_i}{M_0^2 - M_1^2}$$

$$(24.25)$$

where M_1 and M_0 are the chosen lower and upper values of mean wearout life and ΣT_i is the accumulated life lived by r components up to the rth wearout failure. Of course, only wearout failures are counted here.*

The test described by Equation (24.25) is used when the standard deviation of the normal distribution of wearout life is not known. If the standard deviation σ is known, the equation becomes

$$\frac{2\sigma^2 \ln B}{M_0^2 - M_1^2} - \frac{2(M_1 - M_0) \overset{r}{\Sigma} T_i}{M_0^2 - M_1^2} < r < \frac{2\sigma^2 \ln A}{M_0^2 - M_1^2} - \frac{2(M_1 - M_0) \overset{r}{\Sigma} T_i}{M_0^2 - M_1^2}$$

$$(24.26)$$

A test for the standard deviation σ when M is known is given by the following equation:

$$\frac{2 \ln B}{\dfrac{1}{\sigma_0^2} - \dfrac{1}{\sigma_1^2}} + \frac{r \ln \dfrac{\sigma_1^2}{\sigma_0^2}}{\dfrac{1}{\sigma_0^2} - \dfrac{1}{\sigma_1^2}} < \overset{r}{\underset{i=1}{\Sigma}} (T_i - M)^2 < \frac{2 \ln A}{\dfrac{1}{\sigma_0^2} - \dfrac{1}{\sigma_1^2}} + \frac{r \ln \dfrac{\sigma_1^2}{\sigma_0^2}}{\dfrac{1}{\sigma_0^2} - \dfrac{1}{\sigma_1^2}} \quad (24.27)$$

---

* Tests for the mean and for the standard deviation of a normal distribution, and tests for the binomial distribution of non-time-dependent events are described in Wald, *Sequential Analysis* and in A. Hald, *Statistical Theory with Engineering Applications*, John Wiley & Sons, Inc., New York, 1952, chap. 24.

If M is unknown, the sum $\sum_{i=1}^{r} (T_i - M)^2$ is replaced by the sum of squares of the deviations from the sample mean, i.e.,

$$\sum_{i=1}^{r} T_i^2 - \frac{\left(\sum_{i=1}^{r} T_i\right)^2}{r}$$

A binomial reliability test, with chosen reliability values $R_0 > R_1$, is conducted as follows:

$$\frac{\ln B}{\ln \dfrac{R_0(1 - R_1)}{R_1(1 - R_0)}} + \frac{N \ln \left(\dfrac{R_0}{R_1}\right)}{\ln \dfrac{R_0(1 - R_1)}{R_1(1 - R_0)}} < F < \frac{\ln A}{\ln \dfrac{R_0(1 - R_1)}{R_1(1 - R_0)}} + \frac{N \ln \left(\dfrac{R_0}{R_1}\right)}{\ln \dfrac{R_0(1 - R_1)}{R_1(1 - R_0)}}$$

$$(24.28)$$

where F is the number of unsuccessful trials (failures) out of a total of $N = S + F$ trials (S is the number of successful trials). The number of failures F is entered on the ordinate, and the total number of trials N on the abscissa. An accept decision is made if R is equal to or larger than R_0, and a reject decision is made if it is smaller than R_1.

To sum up, sequential reliability tests have the advantage that they allow a comparatively early decision as to whether or not the equipment is at least as good as specified. If the answer is positive, we still do not know by how much the equipment is better. In systems we are not necessarily interested in the actual reliability value as long as we have a specified confidence that a minimum acceptable value is exceeded. Reliability measurements which measure the actual reliability are therefore seldom used for complex equipments at the testing stage. The actual reliability can be computed later from field service reports using the principles explained in Chapter 23. With components the situation is different. The knowledge of the actual failure rate of components is important in reliability design. If we knew only that the failure rate of a component type under certain operating conditions is smaller than some numerical value without knowing by how much such components are better (for instance, they could be better by two orders of magnitude), overdesign would necessarily result and the cost of equipment could rapidly increase. For components, therefore, reliability measurements are an invaluable tool. Once the actual reliability of a component type has been established by measurements, subsequent production lots can be tested by means of sequential tests, using as the lower limit m_1, the reciprocal of the actual failure rate established in the first measurements, or a figure close to it. If production deteriorates, sequential tests will discover this.

The sequential tests discussed so far are probability ratio tests. With the conventional risks of ten per cent, these tests result in very high confidence levels when accept decisions are made. For instance, an accept decision in an exponential test is possible earliest when $T = 4.4m_1$ unit-hours are accumulated without failure, amounting to a 98.7 per cent confidence level that the true m is equal to or greater than m_1, since, according to Equation (22.31), the chi-square value is

$$\chi^2_{0.013;2} = 2T/m_1 = 8.8m_1/m_1 = 8.8$$

Such high confidence levels are, however, seldom required, since reliability testing is a time consuming and expensive process. Had we been satisfied with a 70 per cent confidence that the true mean time between failures exceeds the specified minimum m_1, we could have terminated the test with an accept decision as soon as $T = 1.2m_1$ unit-hours were accumulated without failure. Had one failure occurred prior to $1.2m_1$ hours, we would have extended the test to $2.44m_1$ unit-hours, at which time an accept decision would have been possible, since with only one failure $\chi^2_{0.3;4} = 4.88$ for a 70 per cent one-sided confidence level. With a second failure prior to $2.44m_1$ hours, we would have to continue the test until $3.61m_1$ hours accumulate with no third failure.

By following this procedure a sequential test with a constant confidence level can be designed, and substantial time savings can be achieved. Such a reduction of test time is of particular value for projects with little lead time and when the need for speedy development requires rapid decisions. The test can also be truncated at any time T between two failures. The estimate of m, then, is simply T/r, and two-sided confidence limits can be assigned. For instance, had we truncated the test above at $T = 3m_1$ hours, by which time two failures had been counted, the optimum estimate would have been $m = 3m_1/2 = 1.5m_1$. We would then have a 60 per cent confidence that the true m lies somewhere between m_1 and $3.65m_1$ or an 80 per cent confidence that it lies between $0.77m_1$ and $5.7m_1$. This may be very acceptable in most cases, particularly when it can be supported by a thorough reliability analysis of the design indicating the direction in which the test result might have been biased.

Reject decisions in constant confidence sequential tests follow considerations similar to those for accept decisions, except that the lower percentage points of the chi-square distribution are used to establish reject criteria.

Figure 24.2 compares a 90 per cent constant confidence level test (Test 1, shown by the plain lines) with a probability ratio test with $\alpha = \beta = 0.1$ and $m_0 = 2m_1$ (Test 2, shown by the parallel dashed lines). It can be seen that the 90 per cent confidence test already "accepts" without failure at $2.3m_1$. Truncation lines are also shown. Thus, if the test is truncated at $12.5m_1$, Test 1 "accepts" with 12 failures, while Test 2 "rejects" with only 10 failures. However, an accept decision with 12 failures at $12.5m_1$ is natural, since at that point the best estimate of m is $12.5m_1/12 = 1.04m_1$. Actually, we would expect an equipment with $m = m_1$ to fail about 12 times in $12m_1$ hours. To reject such equipment would be the same as, after having tossed a coin 10 times and obtaining 5 heads and 5 tails, making a definite statement that the coin is not balanced. Also note that whenever an accept decision without truncation is made in Test 1, the confidence that the true m exceeds the specified

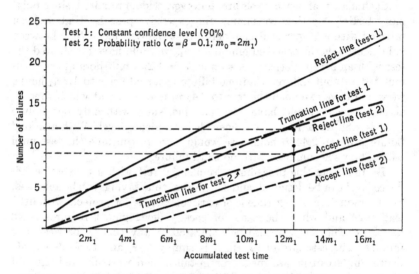

Fig. 24.2. Comparison of different sequential tests.

minimum m_1 remains constant, i.e. 90 per cent. Of course, only exponentially occurring failures must be considered, but wearout tests can also be designed.

 Reliability measurements, as well as all sequential reliability tests, can also be conducted on an accelerated basis, that is, under increased stress conditions. The results obtained from such tests refer, of course, to the accelerated condition and not to normal operating conditions. It becomes necessary to convert realistically the measured parameters into parameters which will prevail under normal operating conditions. To make this conversion, the test acceleration factor F must be known. In general, this factor is different for different types of components as well as for different kinds of stresses. It is defined as the ratio of the value of the investigated parameter, such as the mean time between failures, under a set of accelerated stress conditions to the parameter value under normal operating stress conditions. For instance, in an exponential test

$$F = \frac{m_{normal}}{m_{accelerated}} \qquad (24.29)$$

Thus, if we know from previous controlled tests such as those described in Chapter 15 by Equations (15.4), (15.5), and (15.6) that increasing the test operating voltage, say, 20 per cent over the normal operating voltage reduces the mean time between failures of a particular component population by one-half, which means that the failure rate is doubled, the test acceleration factor is 2. We can then run all subsequent sequential tests on this basis, i.e., by applying 120 per cent voltage and testing for one-half of the specified m_1. It is obvious that accelerated tests can result in tremendous savings in test time and cost, but the utmost care and verification are necessary before any conversion factors can safely be applied.

Accelerated tests, as well as sequential tests, can be adapted to testing for life parameters other than the mean time between failures or mean wearout life. For instance, tests can be designed for the mean number of operating cycles between failures. The parameter time, measured in hours, is then replaced by the number of operating cycles accumulated in the test.

Chapter 25

THE IMPLEMENTATION
OF RELIABILITY

WE HAVE DISCUSSED in detail probability calculations and statistical techniques pertaining to reliability design and testing, but the question of how to implement reliability remains to be answered. Where do we begin and how do we proceed?

A brief look back to the historical beginnings of a systematic approach to the reliability problem is revealing. When the multi-engined aircraft emerged between the First and the Second World War, it was commonplace to perform probability calculations for the propulsion system such as the probability of one out of two engines failing as compared to the probability of one out of four or two out of four engines failing, and to draw conclusions from the results as to the relative probability that the propulsion system of an aircraft would not fail in flights of comparable length. Also, advance information was gained as to the relative amount of engine overhaul maintenance for aircraft of different configurations. These were certainly the beginnings of reliability—interpreting reliability as the probability of surviving an operating period of some given duration under certain environmental operating conditions. But there was no direct attempt to define the reliability concept, and even less to arrive at numerical requirements.

When commercial airlines began to expand their operations, a second aspect of reliability emerged, that of gathering statistical information about the frequency of failures of various equipments in various types of aircraft in the form of average number of replacements; however, the point often missed was that to utilize such information for probability evaluation purposes it was also necessary to compute the total accumulated unit hours of failure-free operation. The replacement statistics, although useful for estimating the need for spare parts, were useless for reliability studies. However, in the thirties, aircraft failure and accident rates began to be compiled statistically on an accumulated-operating-time vs. frequency-of-failure basis, as observed in actual operation over prolonged periods, and the first numerical failure-rate values began to become available for aircraft taken as one system, although still not for individual components.

From these failure rates conclusions were drawn as to what improvements could be expected or what goals should be worked toward. Aircraft reliability and safety levels, as probability concepts, became a reality to which considerable thought was given. Sir A. G. Pugsley in his papers, "Note on Airworthiness Statistics" and "A Philosophy of Aeroplane Strength Factors," published in 1939 and 1942 by the Aeronautical Research Council, London, required that the accident rate of aircraft, "when all causes of failure which can result in an accident are considered," should not exceed 0.00001 per hour, and that in this figure the structure failure rate should not exceed 0.0000001. This can be considered as one of the earliest "specifications" of aircraft safety; we would interpret it today as a requirement for an $R_{safety} = 0.99999$ for one hour of flight, i.e., as a probability of 99.999 per cent that in one flight hour no accident will occur, or that no such failure will develop in one flight hour which could cause an accident.

Obviously, these beginnings were a basis for what gradually developed into the failure-rate addition theorem to be used when computing the reliability of exponential systems from component failure rates, and into the product law of reliability for series systems and the product law of unreliability for parallel systems.

Another historical development in the same direction took place in Germany during the Second World War. Robert Lusser, one of the reliability pioneers, narrates how he and his colleagues, while working with Wernher von Braun on the V1 missile, met with the reliability problem. The first approach they took towards V1 reliability was that a chain cannot be stronger than its weakest link. Thus, the missile will be as reliable as this weakest link can be made, or as strong. Although the V1 was a comparatively simple system, they experienced failure after failure because some component failed in each trial. The missile was 100 per cent unreliable at the beginning in spite of the great efforts made in selecting the components. From the weakest-link concept, which was obviously wrong, they proceeded to the concept that all components must somehow be involved in system reliability, because in some trials it happened that good components, considered to be the strong links in the system, failed and caused system failure. Thus the philosophy emerged that system reliability somehow equals the average of the reliabilities of all the components in the system. But the system was still very much worse than the average component reliability, as found in component tests. No progress was made until one day a mathematician, Erich Pieruschka, who worked with the von Braun team on other problems, was consulted and gave the surprising answer that if the probability of survival of one component is $1/x$, the probability of survival of a system of n such components is $1/(x)^n$. In terms of exponential notations, when the failure rates are constant, we would write today the reliability of one component as $e^{-\lambda t} = 1/(e)^{+\lambda t}$ and the reliability of a system of n such components as $1/(e)^{n\lambda t} = e^{-n\lambda t}$, or in the general case, whether exponential or not, when the reliability of one component is $R = 1/x$ then the reliability of a series system of n such components is $R_{\text{system}} = R^n = (1/x)^n$, which is the same as $1/(x)^n$. From Dr. Pieruschka's advice the reliability formula for series systems emerged which is often called Lusser's product law of reliabilities,

$$R_{\text{system}} = R_1 R_2 R_3 \cdots R_n$$

and which showed that the reliability of the individual components must be very much higher than the system reliability. Therefore, new components of much higher reliabilities were designed and built, and the result was that the V1 achieved a 75 per cent reliability. The V1 missile seems to have been the first airborne system into which, at the later stages of its development, reliability was consciously and successfully designed from specified component reliabilities verified by testing.

A third field in which reliability emerged as a factor of utmost importance is electronics. During the Korean war the unreliability of complex electronic equipment was so disturbing that resolute steps had to be decided upon by the Department of Defense to remedy the situation.* Besides the direct effects on equipment operation, there is another most undesirable aspect of unreliability—unreliable equipment requires a tremendous amount of maintenance. It is reported that even not long ago it cost the Armed Services $2 per year to maintain every dollar's worth of electronic equipment. Or, for an equipment life of 10 years it cost $20,000,000 to maintain every $1,000,000 purchase value of this equipment. It is obvious that the importance of reliability cannot be overemphasized.

* The early efforts to upgrade the reliability of electronic equipment are discussed in a paper by E. J. Nucci, "The Navy Reliability Program and the Designer," *Proceedings of the First National Symposium on Quality Control and Reliability*, New York, November 12–13, 1954, pp. 56–70. In this paper of historical importance Mr. Nucci points out the need to consider reliability in the design stage and defines simplicity, reliability, and maintainability as major design objectives for the electronics industry.

The problem of reliability arises from two main causes. First, there is the complexity of modern electronic and other equipment and systems in which hundreds, thousands, tens of thousands, and even hundreds of thousands of components must operate without malfunction to keep the equipment or system in satisfactory operation for some length of time. Second, there is the trade-off between the magnitude of the safety factors built into the components on the one hand and their weight and size on the other. Thus, the logical approach to higher reliability is to design less complex equipment with fewer components and to use components with higher safety factors built into them. However, this is easier said than done because there are definite performance limitations connected with design simplification, and when components with higher safety factors are used, weight and size are being sacrificed. Still, very often this is the only way to attain higher reliability, and it has become commonplace in reliability design work to sacrifice weight and size by using components at derated levels or components of the next higher rating, which is equivalent to providing for higher safety factors. This, of course, brings up the questions of how meaningful component "ratings" are, how the true strength of components to withstand various stresses is distributed over a lot of components of equal type, how this distribution changes with time and with various stresses actually applied, and what the magnitude is of the maximum operating stresses, especially transient stresses, to which a component can be expected to be exposed in a particular application.

In the implementation of reliability two areas are pre-eminently important—design and production. Next to performance, a certain inherent reliability level is designed into a piece of equipment by the designer. In production, even the slightest carelessness can destroy this reliability. Inspection and quality control guard against it. Ground rules for the designing of equipment for a specified reliability have been given in Chapter 18. The same basic rules also apply when there is no numerical reliability requirement, i.e., when the design is for a qualitative rather than quantitative reliability. A self-imposed reliability goal can be set so that the procedure becomes one of numerical reliability design, or in the absence of a set numerical goal the designer applies the normal techniques of reliability analysis (component derating, etc.) so that he can report the expected reliability level of a new design before a decision is made to go into production.

In Chapter 18 we stressed the importance of a reliability support organization for design, which has to compile and supply to the designer reliable information on the reliability of the components used in design work, including derating information for component operation at various stress levels. When called upon, this group must also provide assistance in analytical reliability work and check the correctness of reliability calculations performed by the designers. It also helps in designing statistical reliability tests. However, it does not and cannot relieve the designer of his responsibility for the reliability level designed into the equipment. Although its capacity is basically advisory, it should have the authority to enforce design corrections necessary to meet reliability goals, if compatible with performance requirements.

In reliability design, the designer's responsibility extends beyond actual design work into the areas of production and quality control. He must on strength of the available reliability data exactly specify components by type and make, specify tolerances, strength of materials, purity of materials, and production procedures. Further, he must specify what reliability tests are required for components and materials to be used in the production of the designed equipment, including component burn-in procedures before equipment assembly and equipment debugging procedures after it leaves the production line. If the specification to which he designs calls for proof of equipment reliability by testing, he also specifies the reliability test procedures for the end product. Considering all these responsibilities, it is obvious that a designer engaged in design work where reliability is important should have a fairly good knowledge of the theory and principles of reliability, and to fulfill his task properly he needs the support of a specialized reliability group.

All that was said above applies also to the design of components, with the exception of a numerical reliability analysis of the design. A numerical reliability analysis is based on the failure rates of the components which the designer integrates into a system to perform a required function. Component failure rates are derived from statistical reliability tests, but the component designer is deprived of the advantage of a numerical reliability analysis during his design work.* His approach to reliability is based on the characteristics and strength of the materials he uses, on safety factors, on reliability testing which supplies information about the causes of component failure, and on taking corrective action where necessary. He can make good use of accelerated testing, physical tests to destruction on a statistical basis, etc., and he specifies reliability measurements and debugging procedures for the production-type components. He also specifies life tests to measure the mean wearout life of components and life standard deviation, and tests to determine the time dependence of the drift of component characteristics. Several of these tests can be combined.

The responsibility for reliability tests rests with quality control, except

* However, in the case of complex mechanical components, such as valves, a preliminary failure rate estimate can be obtained from experience values of the failure rates of basic parts, such as bearings, bushings, shafts, pins, links, seals, etc., and from the safety factors applied in the design of such parts.

for the tests performed on models or prototypes during the developmental stages of a product which are a responsibility of design engineering. Reliability thus adds a new dimension to quality control work without subtracting anything from traditional quality control work and methods. It extends quality control work into the time domain, and greatly increases the areas of activity and responsibility of the quality control organization. It transforms the quality control organization into a quality and reliability control organization.

The question, "What is the difference between quality and reliability?" is often asked. In engineering, traditionally, quality means good performance and longevity. Of course, the same can be said about reliability. However, quality control measures only instantaneous performance and its variations from specimen to specimen by statistical methods to determine whether production satisfies design requirements. When the instantaneous performance parameters of a product, as it leaves the production line, are found to be within specification limits, the product is released for service. Thus, quality control is concerned with the "as is" performance of materials, parts, and products, and with production processes as the product is being built and when it is new, before it accumulates any appreciable amount of operating life.

When reliability is introduced as a parameter, quality control maintains all its original functions, methods, and procedures, only extends the performance measurements from the instantaneous "as is" time-independent domain into the operating-time and life domain. In reliability, not only does it matter what the initial number of defectives is and what the variation of performance characteristics, but it also matters how long a product will maintain its original characteristics when in operation, how the variations spread with time, what percentage of the components will become "defectives" in the first few hours of operation and, therefore, in the debugging period, what the percentage will be in later operation when the debugging period is over and, therefore, what the constant failure rate will be during the useful life, and finally, what the wearout life is and how it is statistically distributed. The statistical techniques used in reliability testing are very similar to those of the conventional quality control methods, with time added as a new dimension. Quality control engineers therefore find it comparatively easy to familiarize themselves with reliability testing procedures and to add reliability testing to their normal activities.

Only when reliability testing is added to conventional quality control methods can the traditional meaning of quality—good performance and longevity—fully emerge, because good quality is characterized by two attributes, the state or performance of the product when it leaves the production line and its capability of maintaining that performance in service. To control this second attribute, quality control must expand into the field of reliability testing and become a quality and reliability control organization.

However, quality control is not directly concerned with reliability design any more than is it concerned with performance design. Reliability design, as pointed out earlier, is the designer's responsibility. It is an entirely different type of work, based on a few probability theorems and their sophisticated use, whereas quality and reliability control work is based on the methods of statistical evaluation. The task of quality and reliability control is to control the production process so that the end products comply with the inherent performance and reliability designed into them by the designer.

The reliability control part of the quality and reliability control program pertaining to production generally covers as a minimum the following areas: (a) verification of reliability tests and debugging procedures performed by the suppliers to contract specifications; (b) performing reliability tests and debugging on incoming parts, materials, and components if no assurance exists about vendor tests; (c) tightening of in-plant inspection in all production areas where reliability is vulnerable (tolerances, potting, avoiding contamination, soldering, connections, correct wiring, riveting, welding, etc.); (d) performing debugging of end products; (e) performing reliability and longevity tests on end products as applicable according to specifications.

Whenever any of these activities shows a drift in an unfavorable direction outside the statistical reliability control limits, that is, when assignable causes of reliability variation exist, action is taken to correct the situation where necessary. An especially useful method of reliability control testing is that of sequential tests, described in Chapter 24.

How an over-all reliability program is organized depends on the size of the company, the type of product, and the production quantity. The organization of the implementation of reliability is usually a good measure of a company's consciousness of the reliability concept and of its awareness of what is involved in reliability.*

Before any serious reliability work can begin, an education in reliability principles, theory, and methods must be offered to all engineering personnel. The education can be tailored to the specific needs of the attending group of engineers. For instance, when design engineers are involved, emphasis is placed on reliability analysis of designs, the pertinent methods of reliability calculations, and reliability design techniques, including maintainability considerations and calculations. For quality control engineers, the emphasis is placed on the statistical techniques of

* For more information on how to establish and manage reliability programs see *Reliability Training Text,* Institute of Radio Engineers, New York, 1960.

reliability measurements and testing, etc. Common to all groups, of course, is education in the basic concepts and principles of reliability. Graduate engineers require about twenty to forty class hours of theory, with examples of reliability problem solving by numerical calculations, to establish a sound background in reliability theory and methods from which they can start to develop their own experience in actual reliability work without making grave mistakes. Extra time is needed by the engineers to solve problems on their own as a kind of "homework" exercise, and to demonstrate how much they have profited from the offered reliability education. As an incentive, engineers who pass a full reliability course successfully should be given special recognition by their company; for instance, a certificate showing the extent of the course covered may be awarded. They have in fact become more valuable staff members, and this should be entered in the personnel records.

Next to reliability education, the organization of a specialized reliability support group to assist in reliability design efforts is required. Besides assisting in complicated cases of design reliability analyses, an important task of such a group is to obtain reliable information on the reliability of all parts and components used in the company, which includes failure rate derating curves, drift vs. time and life characteristics, etc., because without this basic information numerical reliability analyses can be wrong by orders of magnitude, and designing to quantitative reliability requirements becomes a trial-and-error procedure with added expensive testing and redesign work. The group has several ways to compile this vital information about components, such as vendor reliability test reports, field failure data, by its initiating and conducting reliability measurements, and by exchanging information with the industry. The group continuously supplies reliability data sheets on parts and components to the design department as data become available, and data corrections when additional information indicates the need for such corrections. It also monitors reliability calculations made by the designers.

Sometimes the objection is raised that reliability is an expensive proposition. There is no doubt that the expense connected with reliability procedures increases the initial cost of every device, equipment, and system. However, when a manufacturer can lose important customers because his products are not reliable enough, there is no choice other than to incur this expense. How much reliability is worth in a particular case depends on the cost of the system and on the importance of the system's failure-free operation. If a component or equipment failure can cause the loss of a multimillion-dollar system or of human lives, the worth of reliability must be weighed against these factors.

The financial effects of reliability over the designed operating lifetime T of a system can be obtained by comparing two equipments built to perform the same function in the system, one more reliable with a lower failure rate L but with a higher initial cost C_{i_1}, and the other less reliable with a higher failure rate H and a lower initial cost C_{i_2}. Over the time T (design life of the system), the more reliable equipment can be expected to fail TL times, and the less reliable equipment TH times. If the total cost involved in a failure of the more reliable equipment is C_{f_1} and that involved in a failure of the less reliable equipment is C_{f_2}, where the cost C_f includes the cost of replacing and/or repairing the equipment as well as the loss, if any, suffered by the system owing to the equipment's breakdown, the cost of equipment failures over the time T becomes TLC_{f_1} for the more reliable equipment and THC_{f_2} for the less reliable equipment. With the initial costs added, we have to compare $C_{i_1} + TLC_{f_1}$ with $C_{i_2} + THC_{f_2}$ from which we can draw conclusions as to the worth of increasing reliability (i.e., reducing the failure rate of the equipment to L) and as to how much the initial cost C_{i_1} can be increased.

More elaborate cost formulas can be worked out at liberty by splitting up the cost of equipment failure C_f into its component parts, such as cost of equipment spares and repairs (taking into consideration that a repaired equipment is again available as a spare), the financial losses incurred by the system breakdown due to equipment failure according to whether a spare is immediately available or not, the effects of the more reliable equipment if it happens to be heavier or larger (which is not necessarily the case) on system operating cost in fuel or power consumption, payload, etc., and capital amortization and currency devaluation over the time T.

In general, however, it is true that money put into a well-administered reliability program will result in spectacular increases in product reliability which, in the same degree, will reduce the occurrence of failures and therefore the total cost of failures TC_f over the operating life period T. It is well worth-while to pay more initially for a reliable product where reliability is important. In the final analysis, substantial financial savings result for the consumer. For the producer it is a matter of remaining in business. However, his business volume and profits will be substantially increased once his reliability reputation is established. The question, therefore, is not only how much reliability costs, but also how much the lack of reliability costs. On a national scale, the question is how much unreliability the nation and its taxpayers can afford.

Chapter 26

THE STATE-OF-ART
OF RELIABILITY

A BRIEF SUMMARY is given here to elucidate the state-of-art of reliability as presented in this book and to indicate the direction in which more effort is needed to raise the standards of reliability work.

This book was written for the practicing engineer who needs an exposition of the theoretical concepts of reliability and practical formulas to solve reliability problems in design, analysis, and testing in a way consistent with the state-of-art. No claim is made of its being an all-embracing, mathematico-statistical treatment, which is yet to be presented by mathematical statisticians.

Throughout the book emphasis is placed on the derivation of reliability equations, with examples of their use, in order to enable the reader to derive correct formulas for practical applications, which differ from case to case. There is, of course, no such thing as a single, all-purpose reliability formula. However, there exist basic equations, given in this book, which must be applied to obtain sets of formulas for particular problem solutions. The case can be compared with circuit theory where, by applying basic equations, sets of formulas must be derived separately for each specific circuit.

Reliability is based on probability and statistics. As in any statistics and probability calculations, assumptions have to be made, time and again, about the distribution of various kinds of failures which affect reliability. It must be realized that reliability equations approach reality only to the extent that the actual distributions approach the assumed models. Whenever a distribution is assumed, it always remains just a model, whether it is exponential, normal, logarithmico-normal, binomial, chi-square, Weibull, or gamma, etc. Distributions of actual samples never fit exactly. Consequently, when we evaluate actual statistical data, i.e., when we calculate means, deviations, or probabilities on the strength of models, we can never be sure that the calculated values represent the true population. They would approximate it if the assumption about the distribution form were correct. We nevertheless must proceed in this way and use statistical model distributions developed by the science of statistics, and we must aim at the best correlation—that is, we must choose that form of distribution which appears to be closest to a given sample.

In reliability we make estimates and predictions by making correct use of scientific procedures. However, we have to realize that they are just what these words imply, and nothing more. They can be regarded as good estimates and predictions as long as they are not disproved. When the science of statistics comes up with better-fitting distributions, we shall gladly accept them.

Besides the problem of statistical models, in reliability we have the problem of ever-changing probabilities. We know that when a component exhibits a certain probability of survival or failure rate under one set of environmental and operating stress conditions, these parameters change immediately with even the slightest changes in the stress conditions. Changes in environment are often quite drastic even in the operation of one and the same system, and even more drastic from system to system. It is in this field that progress is most needed. What are the laws governing changes of failure rates with temperature, voltage, current, and other stresses?—that is a most important question.

This is where the physicists may enter the reliability arena. They have quite a difficult problem to tackle which, when satisfactorily solved, will save a lot of testing and will greatly increase the precision of reliability analyses and predictions. Correct failure rate transfer curves can now be obtained only by testing at several stress levels, and then fitting the curves without having scientifically derived and proven models available.

There can be only little doubt that the problem of component failure rates is the most pressing one in the entire field of reliability. It is often quite impossible to get information on component failure rates, even from renowned component manufacturers, although very good testing methods are available. Nowadays everybody is selling "high reliability" components, at high prices, but when you ask their failure rate, at least the nominal failure rate when operated at rated stress, no answer can be obtained—that is, no reliability tests were made. However, there are a few laudable exceptions, and manufacturers who are reliability-testing their components and making proven failure rate and failure rate derating figures available should receive first attention from prospective buyers.

There is also the problem of weeding out early failures from component lots. Traditional quality control sampling and testing methods were limited to measuring the proportion of initially defective components. In reliability there is a need to extend quality control operations to the first hours of component operating life, by use of burn-in techniques and statistical models, so as to get rid of not only all initially defective components but also those culprits which cause the early failures.

Another vulnerable point is still with us, which again hinges on components. It is the question of how long components can be operated at certain stress levels before they begin to show wearout effects which increase their failure rates. What is component wearout life and how is it distributed over a lot? Some components also display drift of characteristic parameters. How does this drift proceed with time?

This information is needed for designing reliable equipment, especially when long equipment life is required and high reliability is to be maintained throughout that life. To allow components to wear out until failure occurs drastically reduces system reliability. However, preventive replacement or overhaul can be scheduled economically only if a knowledge of wearout and drift characteristics as a function of component life is available. The need for this information is urgent because these characteristics have a great impact on the reliability, maintainability, availability, and dependability of equipment and systems.

When component reliability characteristics are known, it is possible to predict very realistically equipment and system reliability by analytical techniques and to design exactly to quantitative requirements, without what is often called "intelligent guesswork."* The probability theorems for this type of work are perfect—there is no problem in this field.

In this book the known mathematical tools for reliability analysis and testing are presented, and the state-of-art has been advanced in quite a few areas. Starting from the definition of reliability, the case of exponential chance failures, normal wearout failures and early failures, and their combined effect on component life and probability of survival is presented in detail. It is shown how system reliabilities are calculated in series, parallel, and stand-by, and where Bayes' theorem can be successfully applied.

Component modes of failure and the combined effects of various modes of failure are discussed along with the reliabilities of switching arrangements. The reliability of components operating at various system stress

---

* Quantitative reliability analyses based on an "intelligent guessing" of component failure rates are frequently wrong by orders of magnitude. Therefore, quoted failure rates which are not supported by verifiable test results must be taken with utmost reservation, especially when competitive proposals are being compared.

levels and in various environments is closely considered and it is shown how the knowledge about components is applied in system reliability design work, what the benefits of component derating and redundancy are in complex systems, and how best use can be made of them in design. It is also shown how system reliability and safety benefit from preventive replacement and overhaul policies and what the effects of choice of replacement and overhaul times are. System safety, as a probability concept, is discussed with particular reference to aircraft.

Maintainability, its effects on system availability or operational readiness, and system dependability are explained and workable formulas are derived for these concepts.

Finally, the statistical techniques of reliability measurement and testing are treated in detail for components and systems, with a special chapter on sequential testing covering the exponential, normal, and binomial cases. Accelerated tests are discussed in their proper context of failure rate variations at changing stress levels. A separate chapter is devoted to confidence limit calculations.

Remarks about the role of the quality control organization and of specialized reliability groups in an over-all reliability program and a simple cost model of reliability are included in the chapter devoted to the subject of reliability implementation.

INDEX

287